For Angela,

Best wishes

Marie K. Eder

Leading the
Narrative
The Case for Strategic
Communication

MARI K. EDER

NAVAL INSTITUTE PRESS

Annapolis, Maryland

Naval Institute Press
291 Wood Road
Annapolis, MD 21402

Library of Congress Cataloging-in-Publication Data
Eder, Mari K.
  Leading the narrative : the case for strategic communication / Mari K. Eder.
    p. cm.
  Includes bibliographical references and index.
  ISBN 978-1-61251-047-7 (hardcover : alk. paper) 1. Armed forces and mass media—
United States. 2. Communication policy—United States. 3. United States—Armed
Forces—Public relations. I. Title.
  P96.A752U55 2011
  355.3'420973--dc23

                                                              2011024316

♾ This paper meets the requirements of ANSI/NISO z39.48-1992 (Permanence of Paper).

Printed in the United States of America on acid-free paper

19 18 17 16 15 14 13 12 11   9 8 7 6 5 4 3 2 1
First printing

For the Public Affairs Officers

Contents

Foreword

I didn't start out wanting to be a teacher. I wanted to be a naval aviator. I had wanted to fly since I was eight years old, and after working my tail off, I finally was awarded those "wings of gold" in 1976. Later, when I was medically grounded and transferred into public affairs, I still didn't have any thoughts about being a teacher. Mel Sharpe, my longtime mentor and a professor at Ball State University, where I received my master's degree in public relations, told me that even if I didn't become a teacher, I'd nevertheless spend a lot of time educating senior leadership about the role of public relations. I took notes, but I still didn't think I'd ever be a teacher.

Late in my naval career I realized that teaching was indeed a large part of my job, even though it wasn't in my job description. I realized that much of what I'd been doing throughout my career was explaining to my bosses—and sometimes teaching them—how we could effectively and successfully engage the news media.

Many flag and general officers clearly understood the need to communicate with the media with authenticity and transparency. Working with these senior leaders, we were able to focus on how to strategically communicate our message and help the media tell our story. We often enjoyed the tremendous synergy that resulted from coordinating our message across all channels of public information.

It is this synergy and need to strategically communicate that Major General Eder addresses in this seminal work. She covers the gamut, from a deep philosophical and theoretical discussion on what strategic communication is and should be to very practical, applied tips and techniques commanders, executives, managers, and leaders at all levels can use to successfully communicate and listen. In the end Eder leaves no doubt that strategic communication is not only a function of command but also an essential skill in governance and public diplomacy, especially in the global fight for dominance in the information battle space.

What I find particularly insightful is her vision and road map for the strategic communication force of the future. It is my fervent hope that as you read this book, you will see the logic of her argument. Further, I hope that a long and involved discussion of strategic communication will ensue, a discussion that begins with the assumption for the need to tear down silos and manage the nation's public information capabilities in a more cogent and cohesive manner, just as Eder suggests.

Whether you're a public relations professional, a military commander, or a business executive, this book is an important resource. It provides practical, easily applied tips and techniques for successfully engaging the media as well as an important voice in the discourse on how our nation can transform our communication efforts to allow the United States to fully and successfully occupy the information domain. It's time we took back the high ground.

—*Capt. Robert S. Pritchard, USN (Ret.), APR, Fellow PRSA*

Preface

All communication informs. All communication conducted with intent does more than merely inform. It educates, reveals, restricts, and can elicit strong emotion. Most important, information as an element of national power also influences and can powerfully inform governments, direct public opinion, affect international relations, result in military action, and build or deny support. When intent is tied to strategic purpose, information becomes an incredibly powerful tool in advancing a national or operational agenda. Absent strategic intent, it can become a dangerous, often unfocused weapon.

This book is a collection of essays on various aspects of information and its uses. The common thread is that of strategic purpose, even in crisis, even in seemingly benign settings. The necessity for conscious thought and planning in the application of communication is clear.

No other nation has the technological capability of the United States, no other department in the federal government is as capable of planning as the Department of Defense. No other organization has as much capacity for public information, particularly in the number of public affairs units and the quality of public affairs professionals. It is time for these often unrecognized yet highly talented professionals to receive the recognition that is long overdue. Whether serving in billets that support information operations campaigns, public diplomacy, military information support operations, or public affairs, all of them share some elements of a common skill set in the uses and conduct of the communications process—whether to listen and learn or to educate, inform, or influence—as well as an unflinching commitment to the truth. This book is for them.

Acknowledgments

For Mrs. Kay Ryall Miller, my first journalism teacher and Edinboro University's best writing professor; Col. Bill Smullen, my first hero and role model in public affairs; and then-major Glenn Devier, because he gave me the opportunity to participate.

For my great U.S. Navy bosses, whose professionalism and ability to finesse the toughest issues was a tremendous inspiration: Capt. Tim Taylor, Capt. Robert "Pritch" Pritchard, Capt. Brian Cullin, Capt. Jim Mitchell, and Capt. Chuck Stowe.

For all the sergeants major who taught me not only how to do my job as a public affairs officer but also what my job was: MSgt. Joe Covolo, Sgt. Maj. Paul Turk, Sgt. Maj. Rich Czizik, Sgt. Maj. Phil Prater, Company Sgt. Maj. Ray Cordell, and Sgt. Maj. Patti Winebrenner.

For the members of the Conference Board's Council for Corporate Communications for your acceptance and support, and for all the professionals who work at Fleishman-Hillard Public Relations—you taught me that I could be a real member of the team.

Finally, my thanks go to Mary Pat Begin Ortiz and Fred Ortiz, because your friendship carried me this far; Kristene Greer Walley, because on all levels, you understand; and Debra Hall, for your total support.

Leading the
Narrative

Military Media Relations

Now who has questions for my answers?
—*Secretary of State Henry Kissinger*

♦ ♦ ♦

By the time they have reached the rank of major or lieutenant commander, many military officers have had experience in working with the civilian news media. Invariably those who have had limited experience, or done only one or two interviews in conjunction with deployments, describe their experiences in dealing with reporters as good. Typically I've found in asking officers this question in a media relations setting that the response between a favorable or positive experience is only slightly outweighed by those who feel that the reporter was "out to get a negative story" or "wanted to make me look bad." Invariably, however, those officers are also uncomfortable with the idea of dealing with reporters, fear making a mistake due to the impact on their operation and, of course, their careers, and are generally intimidated by the power of the press.[1]

In a famous clip from the 1970s television situation comedy *The Bob Newhart Show*, Robert Hartley, a Chicago psychiatrist, is interviewed on a Chicago morning talk show. Newhart plays a somewhat naïve guest for the show; he is pleased that he will be on television and appears to be enjoying the notion of impending fame. The interviewer, however, is adept at the type of questions that skewer unsuspecting guests. The clip is amusing, and those watching both laugh at Bob's discomfort and tend to identify with it. He obviously has no game plan or messaging strategy and fails miserably in his efforts to defend the values or successes of his profession. As the clip ends, Bob leans over to the interviewer and with unfailing politeness bids farewell. "Thank you so much for coming. I enjoyed it," she replies. "You would have enjoyed Pearl Harbor," he says.

This clip lasts approximately four minutes and can seem long when viewed by larger groups. Yet it makes the point very well that even the most inexperienced interviewee will relax several moments into a tough interview and then be ready to make his salient points. The stage fright will be past, he is no longer distracted by the camera, and he will be ready to speak. By then, unfortunately, it is almost too late. The comic nature of this introduction makes a great point to those preparing to work with the news media: Don't be like Bob.

That does not mean that military commanders and staff officers can ignore the media's requests for information or their appetite for material. I have often heard reporters say to a spokesperson, "You can choose to be part of this process or not. But I can and will get this story; there is always someone who will be willing to speak."

Reticence for fear of making a mistake is often unfounded, and with good preparation, every media encounter can be a terrific opportunity to use the power of the media and their technology not to just reach out but also to connect with and engage an audience. Good media relations pay off in myriad ways. It builds relationships with reporters and news organizations, reaches through media outlets to connect with audiences, and can result in greater public understanding, support, and good will. Commanders and staff officers should care about information engagement. Preparation and a game plan can make all the difference if the speaker is credible, displays confidence, and is in control of her- or himself and of what she or he plans to say during the process.

A great deal of discussion in the summer of 2010 focused on the topic of media access to commanders, particularly within the military public affairs (PA) community. Many feared the impact of the *Rolling Stone* article and the subsequent firing of Gen. Stanley McChrystal would make commanders more reluctant to engage. Secretary of Defense Robert Gates reinforced the basics of military interaction with the media in his memorandum on media engagement, stating, "I have said many times that we must strive to be as open, accessible and transparent as possible. . . . We have far too many people talking to the media outside of channels, sometimes providing information which is simply incorrect, out of proper context, unauthorized, or uninformed by the perspective of those who are most knowledgeable about and accountable for inter- and intra-agency policy processes, operations, and activities."[2]

There is some evidence that the incident may have had an impact on the military's reputation as one of America's respected institutions. The annual Gallup poll "Confidence in Institutions," published 22 July 2010, revealed a definite downturn: " 'The military,' at 76% again tops the list as it has since 1998, but is down six points from 2009. Gallup does not offer thoughts on the reasons for the drop, but it is plausible to believe that the General McChrystal–*Rolling Stone* episode played some role for those answering the 8–11 July survey. In March's Harris counterpart poll (measuring confidence in *leaders* of institutions), the military was up one point and number one on its list. Gallup's headline was 'Congress ranks last in confidence in institutions.' Gallup states it will follow up with additional data soon."[3]

A first examination of what constitutes the reputation of the military in the public mind must take into consideration a number of influencers.[4] There is the public image, one that is influenced by understanding of the military's role in fulfilling the national political will, the image of how that Army or force acts or comports itself in conflict or peacekeeping missions, and the local views of

Confidence in Institutions—% Great Deal/Quite a Lot			
Now I am going to read you a list of institutions in American society. Please tell me how much confidence you, yourself, have in each one —a great deal, quite a lot, some, or very little?			
	June 14-17, 2009	July 8-11, 2010	Change
RANKED BY 2010	%	%	
The Military	82	76	-6
Small business	67	66	-1
The police	59	59	0
The church or organized religion	52	48	-4
The medical system	36	40	4
The U.S. Supreme Court	39	36	-3
The presidency	51	36	-15
The public schools	38	34	-4
The criminal justice system	28	27	-1
Newspapers	25	25	0
Banks	22	23	1
Television news	23	22	-1
Organized labor	19	20	1
Big business	16	19	3
Health Maintenance Organizations (HMOs)	18	19	1
Congress	17	11	-6
GALLUP			

FIGURE 1-1. Chart from 2010 Gallup poll, "Confidence in Institutions."

service members and their families in their communities, and there is also the role of external champions or spokesmen. These can range from sports figures that support the military to former service members (with both good and bad experiences to color their views) to retired generals, commentators, and many others.

News Overload

As much as these segments of society influence the public view of the military, media coverage of military operations are likewise influenced by the changing nature of the news media itself. There is what reporters term "a public hunger for information" as justification for their inquiries. When reporters talk about the public's "right to know," our response should always be initially, "The right to know what?" This so-called hunger is the yin of the twenty-four-hour news cycle. The yang is the need to fill space within that cycle with information and content. Influencers include the downsizing of major news organizations over the past ten years, including international organizations, with a resulting lessening of international news coverage. There is more of an internal focus, and given the international economic crises of the past several years, a greater internal focus on national and local domestic agendas.

The media may be more fragmented than ever, but there are also more reporters now than there have ever been before, and coverage of military operations and soldiers presents major opportunities, whether to tenured and experienced international and national media representatives or to freelancers, book authors, or bloggers. These in turn present growing challenges to commanders, in terms of not only diversity but also sheer numbers: "Fewer than 30 reporters accompanied the entire invasion force into Normandy, France, on 6 June 1944. More than 500 journalists appeared within hours to cover combat operations in Grenada (1983) and Panama (1989). The trend toward larger numbers continued as more than 1600 news media and support personnel were present at the beginning of Operation Desert Storm (1991) and . . . more than 1700 media representatives covered peacekeeping operations in the American sector of Bosnia in 1996. More than 2700 reporters accompanied NATO forces entering Kosovo in 1999[5] and in the nearly ten years of combat operations in Afghanistan and Iraq, the numbers easily exceed all those of previous conflict coverage."

Audiences are ever more fragmented, with a greater variety of niche media exerting their appeal, and the growing focus on blended news and entertainment can likewise have a startling if not negative effect on coverage of military operations. Another effect is that of "sound bite journalism." Many officers are shocked to learn that even as they prepare for in-depth media interviews, their time on camera or their quotes in print will be extremely brief. Their expectations generally range from one to two minutes of air time, but the reality is more like seven to ten seconds.

Technology is another factor, with both the necessity to check online coverage to the growing pressure to provide an online presence for units and organizations. Many service members have their own blogs, email distribution lists, and websites. Digital cameras record soldier music videos, and those videos find their way to YouTube and other sites where viral celebrities are created daily. Commanders must be aware of the power and reach of the social media phenomenon. Do they support their soldiers who make a parody of a Lady Gaga hit music video or do they discourage these acts in a combat zone? The right answer is sometimes hard to discern.

In terms of news organizations, the largest in the world is the BBC, with over 365 million viewers daily. The twenty-four-hour news cycle, although in broad existence since the mid-1990s, continues to present challenges for military commanders. How can this demand be managed? How can the opportunities this medium presents be used? There is a potential for rapid and real-time video coverage now from every corner of the world and at any time: "The effect of real-time television (and news media reports in general) is directly related to the unity, coherence, and communication of existing policy. If there is a policy vacuum or if officials are searching for a new policy, media reports can have a decided effect."[6]

Government officials are generally becoming more sophisticated in dealing with the pressures of the camera and the persistence of the news cycle. Many senior leaders are adept at managing the interview opportunity to transmit strategic messages and influence future events or outcomes of policy efforts. One instance that resonated with military audiences is the example of then-general Colin Powell in Operation Desert Storm. When Powell said, "We are going to cut it off and we are going to kill it," he was not talking about Saddam Hussein. He was talking *to* him— via CNN. Other examples of the strategic application of policy in media encounters abound. One of the most easily recognizable is the Cold War challenge issued by President Reagan with the famous words, "Mr. Gorbachev, tear down this wall."

Many commanders at the operational level recognize that it may be difficult for them to have such a far-reaching impact using media at their level. However, beyond the potential for media coverage is that of the individual observer who can also video, post, or comment. Anyone can be a reporter. The increase in both risk and opportunity for the military commander is exponential; the number of engagement opportunities can be overwhelming. Many feel the pressure to be better informed, yet with such a great number of sources of information available, many feel less informed than they were even a few short years ago.

Typically, in media relations sessions with commanders, I discuss the topic of bad news. When asked to recall newscasts they have seen in the past several days or news stories they have read, many realize that they tend to recall only negative stories, and even those are ones that have lingered in a week-long news cycle. We discuss the notion that while good news does not last, bad news does. Coverage of bad news can, in some long-term scandals, follow a pattern. This is a pattern that we have seen many times over the years, in coverage of stories that go back to the days of Watergate:

1. The event
2. Initial report of the event
3. Downplay of the event
4. Leaks about the event
5. Public indignation over the event
6. Investigation of the event
7. Repeat steps 3, 4, 5
8. Investigation of the cover-up of the event
9. Repeat steps 3, 4, 5
10. Editorials and commentaries about the event

More recent examples include the year-long international coverage of the reported infidelities of President Bill Clinton. And military examples range from the explosive, long-term coverage of Abu Ghraib prison abuses to stories surrounding the death of former NFL football star and Army Special Forces soldier Pat Tillman and national outrage over treatment of wounded soldiers at Walter Reed Army Medical Center.

The ability of photographs to inflame public sentiment or influence opinion has long been a major factor in negative news reporting. The most egregious photographs or news clips tend to be repeated and repeated until in the public eye a single image becomes synonymous with the event itself, whether or not that is an accurate portrayal.

In yet another informal poll, I asked numbers of officers to relate what they consider in their minds to be the single visual image that is representative of U.S. operations several years ago in Somalia. Inevitably, the response is that officers recall the photograph of the American soldier being dragged through the streets of Mogadishu.

In this light there was a definite possibility that in the first two years following Abu Ghraib, the photographs of prisoner abuse could have fallen into this category and become the ultimate negative symbol of the Iraq War. Since that time, however, there have been hundreds of thousands of diverse images released. With a definite emphasis on outreach, engagement, listening, and involvement, the images of the events at Abu Ghraib prison have been, if not replaced, then supplanted by

President Barack Obama signs the act officially repealing Don't Ask, Don't Tell on 22 December 2010
—*White House photo by Chuck Kennedy*

the sheer volume of photos of soldiers shown building schools, reaching out to Iraqi leaders, posing with smiling children, and providing medical care, clothing, soccer balls, and the symbolic hand stretched out in friendship.

Other Considerations

There are numerous other factors at work in building impressions in the public eye. Various social media and gaming are just a few of them. Blogging is mainstream now, with over 1 million posts a day worldwide. Technorati, which tracks trends and focuses within the "blogosphere," reported in 2009 that there were 133 blog records produced between 2002 and 2009 alone. While the average blog remains a "local" source, with only an average of seventy-five or so readers, nevertheless even the smallest blog has the potential to acquire a million more advocates within a matter of hours or to produce a viral string of incorrect information that can quickly multiply to hundreds of other sites then reproduce again.

Social media is likewise mainstream, with military leaders from the chairman of the Joint Chiefs to company commanders now tweeting about current issues and other topics and service Web pages offering tailored news "pushes" to service members. Military commanders in Iraq and Afghanistan have fan pages with hundreds of uniformed and civilian fans alike, tech-savvy Air Force senior leaders are issued Kindle readers, and Defense Department family programs for service members have websites with the potential for avatars to interact with virtual finance offices and other service providers. There are innumerable ways for units, commanders, and individual soldiers to reach out.

YouTube has already been mentioned. There are pages and pages of soldier-posted videos with rap parodies, and other videos are being added daily. This phenomenon is not just limited to American audiences. There are several well-known karaoke videos done by British army soldiers in Iraq, and there is one remake of the Beach Boys tune "Kokomo" changed to "Kosovo" by an enterprising group of Norwegian soldiers serving with the multinational peacekeeping force in the Balkans. Norwegian officers tell me that this popular video nearly caused an international incident for their forces and engendered an apology by their commanding officers to Kosovo civilian leadership.

I tell commanders that there is no limit to the inventiveness and creativity of their troops. Likewise, there is no limit to their ability to get the story out. As difficult as it may be for commanders to accept this, they do realize that their ability to maintain mission confidentiality is somewhat compromised by the current environment. Whether it is civilians who face job loss due to mission change on one of their installations, family members disgruntled over an extended deployment, or the potential for global reach that one soldier with a camera represents, the need for operational security and information situational awareness is ever more critical.

The Interview

Military officers are likely to encounter news media at various times in their careers. Certainly many career officers in European militaries have their first exposure during a deployment, particularly as company grade commanders. In the American military, I have found it more typical that officers are not exposed to media until they are field grade officers, either commanders or subject-matter experts. I have always counseled commanders to build relationships with community leaders and potential allies and reporters before a crisis occurs. They want to have relationships built before they have to meet a skeptical reporter in a negative or potentially explosive situation. I talk with commanders about the various types of interviews, from the "soft" personal type of a profile report to a crisis response scenario. I provide them with various scenarios in which reporters may badger them or suggest that they have insider information in order to provoke a response.

I cite the case of an Army general who was forced to retire suddenly several years ago. Reporters called the Army Office of the Chief of Public Affairs, Media

Relations Division attempting to gain information about what the general had done to warrant the dismissal. Finally, one reporter asked, "We have information that the former commanding general had an affair with a man. Can you confirm that he did or did not?" We learned later that the reporter did not have any information about an affair. His question, though, had the desired effect. The general, through his advisors, confirmed that he had had an affair with a woman, not a man.

I also talk about the effect of the camera itself. Television interviews can be especially intimidating to someone who has never done one. The camera is large and it can be close. The lights are hot. I often mention the reporter I knew from a Richmond, Virginia, television station who was very lax about doing his homework before conducting an interview. His questions were often rambling or vague. His gift, however, was silence. Mike would stand by the camera and just nod, appearing puzzled. The effect on the interviewees was often visible on their faces.

"Have I not explained myself well? Do I need to restate our position? Give more details?" The urge to fill the silence is often irresistible. Mike would get terrific interview results—and often did not know why.

Apart from the reporter who pretended to have information he did not have, I often caution commanders that they should not expect their view to be the only one that comes out in an interview. They do, however, have the right to ask, "Who else have you spoken with?" and "Where else have you been?" If they are uncomfortable with these questions, I tell commanders to use their public affairs officers to do the background work for them. They should know who the reporters are, their background and writing style if print reporters, their tone in coverage, past articles or news reports, and the time line for the current piece being worked. The public information officer or public affairs officer (PAO) should know the policy position from higher headquarters ("We never discuss plans for future operations," for example) and set the ground rules for the interview, including length of time available and topics to be covered. During a time of possible policy change such as DADT ("Don't Ask, Don't Tell," the policy on homosexuals openly serving in the military), reporters probably will want to discuss these changes and gain a commander's opinion, whether or not it is appropriate.

In media relations sessions we discuss strategies, messaging, and blocking and bridging. In tough or even hostile interviews, a respondent must have a purpose for participating and certain items he wants to discuss as well as points he wants to make. Ten message points may be excessive; one is limiting and too obvious. Three to five are generally most feasible in a half-hour to hour interview session.

Most commanders can relate to these topics, and after discussion, they can then view an interview and pick out message points and how interviewees use them to advantage. One thing that is very important in conducting media interviews is for commanders to understand that they need to have an agenda, stick to their own messages, and realize that an interview is not a military briefing. Practice is important, and preparation cannot be overemphasized.

In classes in which I discuss this topic, I have found that the opportunity to conduct mock interviews creates the greatest anxiety in officers and the most intense focus on doing well. When I teach commanders preparing for deployments, either to a combat zone or to a peacekeeping mission, I find that although they may not enjoy the process, they learn even more from the experience of being before a camera or a group of reporters. Invariably they leave better prepared and more confident in their abilities to tell the stories of their units, their people, and their missions to a broad variety of audiences.

2

Crisis Communication
and Conflict Storytelling

In the information age,
success is not merely the result of whose Army wins,
but also whose story wins.

—Joseph Nye, "Bush's War on Terrorism after the 2006 Election"

◆ ◆ ◆

The information domain is the only battle space that does not exist naturally in nature. The other domains for international diplomacy and warfare, the high ground of both strategic and symbolic significance—the high seas, the skies, and space—exist in the physical world. However, the airwaves, the land of broadband and the Internet, do not. Or do they? In fact, several other domains existed but had to be discovered and navigated, just as we have had to discover the means of travel in outer space, under the seas, and above the clouds.

But the information domain is also a battle for the right of occupancy, the supremacy of occupation, and the fight for the hearts and minds of the populace who live, work, and belong there—as well as those who would. Al Qaeda and other jihadist groups have been enormously successful in using the Internet as a means of occupying that battle space, issuing statements and assertions of fact that largely have gone unchallenged. And in the past five years or so, as their growth has been quietly ignored, the Internet has exploded.

Al Zawraa, a new twenty-four-hour insurgent station established by Al Qaeda in November 2007, was based in Syria and blanketed the Middle East and North Africa. It was established by the Islamic Army of Iraq and quickly became known for broadcasting insurgent attacks against American and coalition forces. While the station closed down in mid-2008 after being jammed repeatedly, resulting in the loss of much of its commercial advertising revenue, the dilemma that the station and others imitating its stance creates cannot be overstated. Such outlets provide ready and effective propaganda for the jihadist cause and recruiting material for insurgency growth efforts. They also provide a valuable source of intelligence for coalition forces and others wishing to learn their methods, aims, and future plans.

The information domain and its challenges and opportunities, however, are much broader than the information needs for dissemination and acquisition realized in the present conflict. This domain presents a challenge for all individuals and government organizations daily, at home and abroad. Occupying terrain in the information domain is critical, not just for organizations, governments, and armies but also for those seeking to understand information and its uses and for those seeking to police and support them. This is especially true in the era of the 24/7 news cycle. The demand for information is high, the space seemingly endless, and opportunities as well as challenges and counterchallenges abound. Here the fight for whose story will win begins. It continues with propaganda, recruiting for various causes and groups, it sells, it pitches, and it conducts damage control. Many voices here compete for audiences and for the attention of the young, the neutral, and those most likely to be susceptible to influence: "The fight transnational terrorist groups have waged has been fought on a virtual as well as a physical battlefield. Indeed, the ideological agenda of transnational terrorism has been conducted almost entirely on the Internet's battlefield."[1]

War is a long-term crisis, one that continues over a long period of time and can have vagaries of support and understanding as well as periods of good and bad coverage. A single crisis is much more manageable in the short term. A single event that results in news coverage of only a day or a full weekly news cycle is more desirable than one that can have a long-term impact on national security objectives or outcomes.

Crisis versus Attack

A crisis invites attention from the news media, from opposition forces, from pundits, and from interest groups and individuals. Whether a crisis occurs in nature, such as a hurricane, tsunami, or forest fire, or is man-made—anything from acts of war to corporate crime—it should not be confused with an attack. An attack, launched deliberately by groups or individuals, can be used either as an attempt to inflame a crisis, enhance an issue, influence outcomes or perceptions, or deflect attention from one element of the crisis to another. Attacks can be very well funded, timed, and sponsored. A crisis is not an opportunity for the defensive organization or individual to respond. One example of a manufactured crisis includes the negative national news reporting on the effectiveness of body armor issued to U.S. troops in Iraq and Afghanistan. To consider this an attack is a matter of perspective; investigative reporting by the news organization appeared to have soldier safety as its primary focus. However, a deeper examination reveals why the government response was limited: The reporter was being given information by competing defense contractors. The contractor who lost the government bid was clearly providing the bulk of the faulty information to the media. To many who were aware of the circumstances, this was clearly a case of war profiteering.

In this context, can the recent Wikileaks fiasco be considered an attack? It looks like an attack, this release of vast amounts of classified information concerning U.S. military operations in Afghanistan. It feels like an attack in terms of result if not intent: assault on the reputation of the U.S. and NATO military organizations, their strategy, and effectiveness. The revelation that much of the raw intelligence data can have devastating effects in terms of release of sensitive names and locations likewise constitutes a real attack. This is true regardless of intent—either by the original leaker or by the organizations posting the material. The real damage here, however, remains to be seen. How will international relationships be affected in the long term? Will America's friendships with allies continue? Time will tell the real damage caused by Wikileaks.

Responses to attacks may be limited and may likewise be limited in effectiveness. It is important for spokespersons to note that in an attack, issues of trust and reputation are at stake. Government spokespersons operating in a long-term crisis environment must recognize that there are a number of factors that may limit their abilities to respond to counterattacks with assured success.

Conflict Storytelling

The first and most significant element for a government spokesperson to realize is that there are stories that appeal to Americans—because of culture, tradition, and precedence. Americans have always enjoyed stories of patriotic service, of determined, creative individuals who succeed against all odds, of captains of industry who start out penniless but rise to great heights through their genius and persistence, stories of teams forged in battle or on fields of play—teams that come from behind to win. These are the inspirational stories of businessmen and presidents, cowboys and soldiers, leaders and eccentrics, millionaires and Olympians. Americans enjoy these stories as they have from the tales of childhood, from Paul Bunyan to Audie Murphy—heroes to be placed on pedestals, admired, and imitated.

What appeals in American stories?

• Patriotism

• The American spirit: never give in, never give up

• Rugged individualism

• Teamwork

• Horatio Alger

• The pedestal

From this standpoint comes the natural perspective that Americans favor the plight of the individual versus the institution and place the burden of proof on the government. Witness the story of Pat Tillman, a professional football player who joined the Army after 9/11 only to die in Afghanistan in 2004 as the result of friendly fire. As is typical in a crisis, there were two stories, that of the event itself and the story of how the event was managed. The tale of his service was cut from the very fabric of the American spirit—a patriotic young man gives up a lucrative sports career for a higher calling, to serve his country and fight its enemies. He was a high-profile soldier and his untimely death naturally generated extensive media coverage. However, the charges of a cover-up and subsequent multiple investigations served to expose the fragile bond of trust that the American people have with their trusted institutions, and in the end the Army's reputation was permanently tarnished. Every year since, and with the publication of several books on his life, the Pat Tillman story plays out once again.

In long-term crisis coverage, the news media often create emotional touchstones to evoke the larger picture of the story and recall its emotional foundations while building dramatic tension anew. Some recognizable examples include the following:

- "This is the one hundredth day of captivity for Americans in Iran."
- "Tomorrow marks the first anniversary of Abu Ghraib."
- "Today is day eighty-nine of the gulf oil crisis."
- "The death toll in Iraq has surpassed one thousand."
- "Ten years ago today Operation Desert Storm began."

These touchstones can be very difficult for organizations to defend against or even counter. For example, there was no so-called anniversary of Abu Ghraib. The date the media referred to was the day the story broke in April 2004. The anniversary the media refer to, therefore, was a construct of their own making.

Crisis Communications Strategy

Do organizations typically have a strategy for response? Ultimately it appears that many do not, other than to merely respond to the question of the day with "talking points" and messages that go to individual charges and accusations, investigative results, and pieces of evidence—not to a larger strategic purpose. In a crisis, particularly a long-term crisis, the importance of having a communications strategy cannot be overstated. A crisis communications strategy can be terrifically effective in enhancing reach, establishing and maintaining clarity of message, and targeting audiences with a specific focus. In an attack in particular, it is very important for an organization to speak "to the choir." Loyal supporters should be continually addressed, and their support must be appreciated and both courted and rewarded.

In terms of technique, an organization, whether government or corporate, should likewise maintain a consistency of spokespersons. Within Department of Defense (DoD) media operations, there are often groups of spokespersons who have structured "accounts." In Army Public Affairs, for example, the Media Relations Division has spokespersons who are divided into teams, each with a specific "beat," ranging from personnel issues to weapons system development.

Providing the documentary evidence is another effective strategy for reducing the media's power in telling one version of a story. Every time an organization, particularly a government organization, releases online the entire report, study, and/or investigation, all of the evidence is there for readers and researchers alike to discover. It reduces the potential for misinterpretation and analysis that could skew both intent and outcomes: "Technology is further shifting power to news makers, and the newest way is through their ability to control the initial accounts of events."[2] It enhances the organization's reputation for transparency and it increases trust.

Timing is another significant factor. Just because an investigation is completed, a report readied, or an announcement approved, does not mean that it should be released immediately. I have likened this to crossing the street without looking left or right for oncoming traffic. While one remains focused on the objective ahead, moving forward without an understanding of the necessity of timing for that move may result in a most unfortunate confluence of events.

Several years ago as the Army completed work on its new field manual for interrogations, it was ready to publish the document and announce its arrival via a public affairs release and launch. Fortunately, someone looked left and right before stepping out into that deadly information traffic snarl and realized that the same day the launch was to take place there was an international conference on torture ongoing in The Hague, Netherlands. Could this unfortunate timing have resulted in negative coverage and a long distasteful trail of commentary in the blogosphere? It could well have done so, but the question was thankfully avoided. Timing really is everything.

Summer news coverage has become in America the domain of the celebrity story. In August, in particular, as Congress recesses, reporters take their own annual holidays with some measure of security that no major story will break in their absence. The stage is left to the stories most easily covered—celebrity drug and sexual excess, crime, and sports speculation. Witness the August stories of years past and their greatest hits: the deaths of Princess Diana, John F. Kennedy Jr., and Michael Jackson; Paris Hilton's antics; murder in Aruba; crash of the Concorde; baseball doping scandals; and more. In the summer of 2010, the political dog days of August appeared to be expanding well back into June. Stories of murder in Peru, actress Lindsey Lohan's jail sentence and struggles with alcohol rehabilitation, the royal wedding in Sweden, and other celebrity stories competed for media space along with stories of crime and politics. The budget crisis in Greece, piracy on the high seas, earthquakes, and forest fires received considerably less coverage. Typically the more serious and less salacious the story, the more it moves gradually toward the background and is forgotten.

Truths and Trends

A number of the factors affecting crisis media coverage today result from the general decline in journalistic standards over the past ten years. While celebrity coverage now appears to be so deeply mainstream as to be unavoidable, it is nevertheless worth commenting upon for its debilitating effects on all other coverage. Every time a public figure proves that he or she is not worthy of remaining on the pedestal of reverence they achieved for themselves, it further deteriorates our faith in those leaders and role models, whether they be Tiger Woods, General McKiernan, or General McChrystal.

A whole generation appears to be tuning it all out. A general audience is ever more difficult for advertisers to reach. The ability to record, select, and eliminate unwanted programs and messages is shrinking audiences for both network news and many television commercials. News is increasingly tailored by, for, and to the individual and serves to meet that individual's needs:

> For the third consecutive year, only digital and cable news saw audiences grow among the key sectors that deliver news. In cable in 2009, those gains were largely captured by one network, Fox, though during the day, a breaking-news time, CNN also gained viewers.
>
> What's more, the data continue to suggest a clear pattern in how Americans gravitate for news: people are increasingly "on demand" consumers, seeking platforms where they can get the news they want when they want it from a variety of sources rather than have to come at appointed times and to only one or two news organizations.[3]

The decline in traditional media, driven as much by the economy in the past several years as by faltering standards, results in ever-fewer reporters chasing larger stories. Numerous topics receive little if any news coverage at all, and big stories grow ever larger.

One positive trend is that communication, via numerous Internet sites and blogs, is becoming ever more interactive and connected. While old and new media remain connected, it is still the agenda-setting function of the mainstream press that drives the comments of blogs and online pundits. Humor is an attendant factor, with comedy shows that criticize media coverage, such as Jon Stewart's *Daily*

Show, gaining in popularity and credence. Comedy has become the watchdog of media that has long been needed.

A growing trend in crisis coverage over the past several years is that of the documentary. While documentaries can result in longer and more in-depth pieces on a certain topic, they can also be agenda-based and reflective of their parent organization's values and perspective. HBO documentaries, for example, tend to be edgy, gritty, and realistic. Documentaries also have a longer shelf life than regular news stories and with repeated showings can continue to garner audiences and attendant critics and supporters.

Crisis Response

Crisis response may be different for government organizations than in the corporate world. The technique of attempting to expose the opposition's strategies can potentially backfire and reflect poorly on a command or a commander. Apologies can likewise work both ways or, if not offered immediately, be disregarded as insincere. Witness the U.S. Army's attempts at public apology in the aftermath of the scandal regarding soldier housing and patient care at Walter Reed Army Medical Center. The efforts were not sufficient to assuage the media's demands for accountability and apology. As renowned author and public relations expert Eric Dezenhall commented, "When outrage takes over, there is no emotion more powerful than the urge to blame."[4]

Damage control takes many forms, from holding leaders accountable, to firing them, to the promise of additional and more in-depth investigations. This is one instance in which the credibility and, ultimately, the "likeability" of the CEO, commander, or spokesperson are important. We tend to trust and believe in those we instinctively like. Direct outreach to specific audiences can be another effective means, not only of reassuring the "choir" but also of building champions from independent sources who can then carry the message further and with greater credibility.

Several years ago the question "Does America support its troops?" was a major topic among senior Army leaders. Many felt that few Americans even realized that the country was at war; with less than 2 percent of the population involved, that was a logical assumption. But then life began to imitate art, or was it the other way around? Veterans organizations began to make it a habit of welcoming home troops at airports and installations across the country; the USO was an active participant in these activities, and airlines followed suit with recognition for service members traveling in uniform. Commercials from beer companies, showing spontaneous applause at airports as returning troops disembarked, followed. And then, as people began to hold their own small ceremonies, spontaneous applause greeted many returning units. Thousands of soldiers have now experienced this phenomenon.

Hundreds of families and friends were crowded inside the Kieschnick Physical Fitness Center on Fort Hood, Texas, for the homecoming ceremony of the 13th Sustainment Command (Expeditionary) on 3 July 2010.
—*U.S. Army photo by Cpl. Jessica Hampton*

Americans like to be part of community events; they like to participate in something special; they want to be able to say, "I was there." Universities across the country welcomed back to school returning troops who were glad to become students once again. Major league sports became proud sponsors of enlistment ceremonies, held at halftime in professional football games or before the start of baseball or basketball games. Typically these events weren't broadcast, but for the hundreds of thousands of fans who witnessed these events and others like them, they too could tell the story: "I was there. I was part of this event; I supported our troops." This sense of belonging goes a long way to helping audiences become supporters of or even members of the "choir."

Conflict storytelling is difficult. It is often highly charged with negative emotion and driven by extreme pressures of time and the harsh spotlight of unrelenting criticism. But the stories being told on an ongoing basis that are ultimately tied to a communications strategy go a long way to mitigating the effects of the bad news day and may open doors to greater understanding and acceptance. This is a critical effort, particularly in the information domain, which our critics and our enemies seek to dominate.

Strategic Communication and
the Battle of Ideas

*I have been commenting on the challenges our country—not just our government—
faces in fighting a war in this new media age. Moreover, while the enemy is
increasingly skillful at manipulating the media and using the tools of
communications to their advantage, it should be noted that we have an advantage as
well: and that is, quite simply, that truth is on our side—and ultimately, truth
wins out. I believe with every bone in my body that free people, exposed to sufficient
information, will, over time, find their way to right decisions.[1]*

• • •

The above quotation and similar comments in recent years have served to
reignite the public debate about strategic communications, propaganda,
and how our government communicates, both at home and to the world.
A great deal of that frustration centers on the existing capability of current Public
Affairs communications structures to deliver the nebulous benefits of "strategic
communications." This situation is not unique to the Department of State, the
Department of Defense, the Army, and other services, or elsewhere in the executive
branch of government. Yet as our government works on transforming to meet
the requirements of a new age, the question of how to transform and strategically
develop communications is one of great concern.

At issue is the concern that America does not communicate clearly with the
world. It often seems that the U.S. government sends "mixed messages" or fails
to clearly and consistently communicate policy. While this has the potential to
frustrate allies and confuse both potential friends and enemies, it also conveys
weakness in the national will to any nation seeking to understand the intent of the
United States with regard to international relations.

In March 2006 Under Secretary of State for Public Diplomacy and Public
Affairs Karen Hughes gave a speech on transformational public diplomacy at the
Baker Institute for Public Policy. In her remarks she talked about six key areas in
which transformation is fundamentally changing the way the State Department
does business. She first discussed how the department is increasing funding for
programs that are working. In particular she mentioned international exchange
programs, a direct form of community outreach, albeit on a global scale. She noted,

"People who come here see America, make up their own minds about us and almost always go home with a different and much more positive view of our country."[2]

Hughes went on to discuss the State Department's emerging strategy concerning public communications. While acknowledging the rapidity of global communications, she touted the department's new Rapid Response Center—not a completely new concept but a hybrid based on the successful model used by Defense Public Affairs during the kinetic phases of the recent wars in Afghanistan and Iraq. The center monitors daily communications worldwide and provides a summary to diplomatic outposts along with America's message in response. This information enables American government representatives to be more effective advocates for U.S. policy. Additionally the establishment of regional hubs to position spokespersons in key media centers like Dubai will ensure even greater presence and reach. Hughes has likewise given ambassadors and foreign service officers greater freedom to reach out, both directly and through the civilian news media.

Finally, Hughes said that State is placing greater emphasis on using public diplomacy to shape policy. From her travels she learned that America hasn't always shaped programs to make their benefits clear to average people. "He's [the president] now instructed us to look at ways to make our programs more effective," she said, "to set clearer goals, focus our programs and partner with the private sector . . . then make sure we communicate what we are doing—a perfect example of the intersection of public diplomacy and policy."[3]

Defense Communications Strategy

In his speech to the Council on Foreign Relations, former Secretary of Defense Donald Rumsfeld commented on the Defense Department's view of the way ahead:

> *Government public affairs and public diplomacy efforts must reorient staffing, schedules and culture to engage the full range of media that are having such an impact today.*
>
> *Our U.S. Central Command (CENTCOM), for example, has launched an online communications effort that includes electronic news updates and a links campaign that has resulted in several hundred blogs receiving and publishing CENTCOM content.*
>
> *The U.S. government will have to develop the institutional capability to anticipate and act within the same news cycle. That will require instituting 24-hour press operation centers, elevating Internet operations and other channels of communications to the equal status of traditional 20th century press relations. It will result in much less reliance on the traditional print press, just as the publics of the United States and the world are relying less on newspapers as their principal source of information.*

And it will require attracting more experts in these areas from the private sector to government service. . . . We need to consider the possibility of new organizations and programs that can serve a similarly valuable role in the War on Terror in this new century. . . . There is no guidebook—no roadmap—to tell our hard-working folks what to do to meet these new challenges.[4]

DoD efforts to focus on the need to improve public affairs were brought to the forefront in 2004 during a "Tank Brief" to the service chiefs of staff on the subject of public affairs. That session was held as the result of a continuing debate centering on commanders' frustration with a communications process that had been not only ill-defined but also little understood.

The strategic communication process has had a long, hard road to develop over the past several years. In 2010 Adm. James G. Stavridis commented on the importance of strategic communication to him as the commander of U.S. Southern Command from 2007 to 2009. He said, "Strategic communication is the ultimate team sport—it must be done as part of a joint, interagency, and commercial system."[5] This is a tremendously insightful view by a senior commander. Even so the term "strategic communication" remains potentially the most misused and misunderstood term in the military lexicon.

In 2004 the DoD began to explore growing a strategic communications capability and structure, supported by the findings of the Quadrennial Defense Review (QDR). The newly established position of deputy assistant secretary of defense (joint communication) (DASD[JC]) was created in December 2005 recognizing the importance of applying strategy to communication. This billet was established to "shape DoD-wide processes, policy, doctrine, organization and training of the primary communication supporting capabilities of the department. These include public affairs, defense support for public diplomacy, visual information, and information operations including psychological operations."[6] The Terms of Reference established for the creation of this position state that it exists to maximize DoD's capability to communicate in an aggressive and synchronized manner. It clearly represents the first formal recognition of the need for a military communication advocate at the highest level.

One of the primary tasks of the DASD(JC) is to drive communications transformation in the department and to implement decisions from the 2006 QDR to improve all aspects of strategic communications. A working roadmap is being developed to provide strategic direction, objectives, milestones, and metrics for success. Just as important, the roadmap identifies program and budget implications of strategic communications initiatives.[7] There are three overarching objectives the roadmap seeks to achieve:

1. Define roles and develop strategic communications doctrine for the primary communication supporting capabilities: public affairs, information operations, military diplomacy, and defense support to public diplomacy.

2. Resource, organize, train, and equip the DoD's primary communication support capabilities.

3. Institutionalize a DoD process in which strategic communication is incorporated in the development of strategic policy, planning, and execution.

There has never been a validated joint requirement for public affairs. No requirement had been established for a PA capability to support joint/combined/ expeditionary operations. The consequences of this omission set the groundwork for failure in communicating operations that developed rapidly and on the global media stage. What commanders expect/want is not described in any detailed fashion, so the services were left to conduct support estimates through their own doctrinal processes. There should be no surprise that the resulting capabilities did not match demands or expectations.

Along with the establishment of the position of the joint communications deputy, DoD took steps to formally assign responsibility for communication proponency to establish a joint structure to provide a rapidly deployable communications capability and to build a capacity to develop both communications doctrine and materiel. These capabilities were embedded in the mission set and function of the Joint Forces Command–based Joint Public Affairs Support Element (JPASE). The evolving JPASE organization, as part of the Joint Forces Command's rapidly deployable Joint Enabling Capabilities Command (JECC), exists to support the integration of communications into warfighter training, to develop operational public communications programs and policies to support the warfighter, and to provide the combatant commander with a rapidly deployable military PA capability—at the beginning of an operation, when public communications are most critical and have the potential to be most effective.

In the past several years, much discussion in the Army has centered on the inability of the existing public affairs structure to serve the Army with a strategic communications capability. In fact, the function had not been empowered and has been barely resourced to succeed. Despite repeated recommendations from studies such as the McCormick Foundation's report *America's Team: The Odd Couple—A Report on the Relationship Between the Media and the Military* following the Gulf War,[8] the Army did not prioritize the public affairs field with the resources necessary for it to serve as the information combat force multiplier it can and should be.

Journalist Richard Halloran explained it this way more than fifteen years ago: "The most important element in the relationship between a journalist and a PAO [public affairs officer] is the policy of the PAO's commander. A commander with an open attitude communicates that tone to his subordinates and enables the PAO to do his job. A commander who wants a palace guard will get it, and with it, most likely, a bundle of bad press clippings. . . . Equally important, when things beyond the PAO's reach go wrong, and they will, the commander must protect him against the wrath from above, just as he would protect another staff officer."[9]

The Army Public Affairs field not only failed to improve in the years following the first Gulf War, but its stature even declined. How did this happen to a career field that seemed to be advancing well as recently as a few years ago? It happened surprisingly in plain view—of Army leaders, public affairs practitioners, and the audiences the Army serves. It happened despite a plethora of studies on the "military media relationship," although nearly all of these deal with the relationship between military leaders and the media. Very few ever address the actual communications business of public affairs or the public affairs professionals who facilitate relationships on both sides of issues.

The balance may have changed as the role of Information Operations (IO) began to rise and gain influence and recognition, at the expense of the less-well-funded and operationally regarded Army Public Affairs organization. This occurred concurrently with the advent of the term "strategic communications" (stratcom) and its subsequent growth in appeal and stature. It seems that one reason for the appeal of both IO and stratcom is the inherent nature of the one-way communications use the terms invoke. Many senior Army operators, as they have historically done, don't trust the press and, by association, similarly distrust their press officers. And while some believe IO, by its very nature, doesn't necessarily require or involve interaction with Public Affairs or the media, it is absolutely essential that PAOs have complete access to, and situational awareness of, any communication interaction in the global information environment. It can be, after all, the most seemingly insignificant communication that can have international or strategic consequences.

Even as the Quadrennial Defense Review addressed the need to implement a culture of strategic communications within the Department of Defense via the Strategic Communications Execution Roadmap, the services were beginning to move forward to make sense of a concept that has been broadly but poorly defined and often little understood. In the Army the concept of developing a strategic communications process was initiated in 2004 with the establishment of a strategic communications team within the office of the director of the Army Staff.

While the team's charter required linking communications to Army strategy and priority programs, it has taken nearly two years to mature the effort to the level that can best be described as "walk" in the "crawl, walk, run" paradigm. Since then the responsibility for all Army strategic communications planning has transferred to the Office of the Chief of Public Affairs, along with the attendant staffing and funding for contract support. Using an enterprise approach to communications across the Army, the new staff began to understand and define their charter; develop relationships with Headquarters strategists, subject-matter experts, and other communicators; and create the structure, processes, culture, and image to communicate the Army's story. Through the Strategic Communications Coordination Group, they moved to develop plans and associated products, such as the Army Communications Guide, furthering understanding of significant Army themes and messages, campaigns, and events to a variety of audiences.

Today there is growing senior staff–level support for the application of strategy to communications and acceptance of collaborative planning processes in crafting major communications campaigns. This initial framework for public affairs is serving as a sense-making device, a construct that allows us to make sense of a new idea.

The progress to date cannot be described as grand strategy on the national level, or even a Department of Defense–level application of strategic communications. The impact of strategic communications planning and processes at the Department of the Army is that strategic communications has become well nested in the Army's strategy for transformation and solidly linked to the National Military Strategy. This is significant. By beginning the hard, detailed, day-to-day work of establishing coordination and development/design processes for communications planning, first at Headquarters and, in the next year, throughout the Army's subordinate commands, the Army has taken the initial difficult steps of building an understanding of what strategic communications is and how strategic communications planning can work.

These efforts have already paid dividends in linking communications to the Army's long-term programs and processes in supporting transformation. As national concepts of strategic communications planning mature and the Department of Defense implementation of strategic communications processes evolve, the Army's efforts to date will ensure the Army is ready to support and compliment those efforts.

Larry DiRita, former special assistant to the secretary of defense, said the headache of undertaking transformation is well worth it: "The old-fashioned idea that you develop the policy and then pitch it over the transom to the communicator is over. You're continually thinking about communication throughout the course of the policy development process."[10]

The Public Affairs Officer

At the unified commands, public affairs capabilities had been historically diminished through restrictions in force and grade structure. A colonel/captain–level public affairs officer serving on the unified commander's staff absolutely cannot compete on a level playing field with the two-star J-3s and J-4s for the commander's time and attention. The senior communicator on a four-star combatant commander's staff must be, at a minimum, a one-star flag officer. The message otherwise is that the communications function is significantly less important than the other command and staff functions.

An effort to remedy this situation through a proposal for brevet promotions has not advanced over the past several years at DoD but shows promise for the future. Recommendations supporting this change first surfaced *over fifteen years ago*, and while the recommendations have great merit, they have languished in a zero-growth environment as being "just too hard" to accomplish.

In 1995 the Freedom Forum First Amendment Center's report, *America's Team: The Odd Couple*, focused on the relationship between the media and the military. The study was extensive and the recommendations detailed and exacting. The report recognized the need for strategic public affairs leadership at the unified commands, stating, "In major conflicts such as Desert Storm, the Secretary of Defense and the Chairman of the Joint Chiefs of Staff should consider assigning an officer of flag or general rank in the combat theater to coordinate the news media aspects of the operation under the commander of U.S. military forces."[11]

This did occur at U.S. Central Command in the early days of Operation Iraqi Freedom. As operations in the Central Command theater began to generate operational velocity on the international stage, it became apparent the Public Affairs colonel did not have the staff muscle to serve the command at that required level. Rear Adm. Craig Quigley, a career public affairs officer, was detailed from the Office of the Secretary of Defense (Public Affairs) to Central Command to serve as its director of Public Affairs. Upon his retirement, Jim Wilkinson, a White House appointee with GO-commensurate rank, was assigned to take his place. When Wilkinson left at the conclusion of major ground combat operations, U.S. Central Command continued to look for a civilian of his stature, experience, and connections to take his place. That search was unsuccessful and the Central Command Public Affairs effort slowly began to revert back to its prewar configuration and capability.

By the summer of 2004, U.S. Central Command's public affairs staff complexion had changed drastically from what it was at the height of the conflict. From a staff of seventy, headed by a general officer or civilian equivalent, to a staff of barely ten, the office remained functional despite the split operations between Tampa and Doha. Obviously such a limited staff was unable to deal with the tempo of communications requirements (either with American or international audiences), which had increased since the end of the conflict. This was not due to a lack of proficiency on the part of the staff but was a direct result of the immense nature of the continuing demands of the global information environment.

Information Operations began to expand to fill that void, although later the overlap in mission sets was largely resolved with an expanded staff in the Public Affairs Office. Public Affairs generated a strategic communications approach to reaching American, allied, and Iraqi audiences and initiated an aggressive communications outreach focus.

The Army's position is that all general officers are both senior leaders and senior communicators. The Army focuses on the need to broaden the baseline communications skills of all Army officers and make them all communicators. Those who choose the Public Affairs Functional Area career path must understand this reality. Following DoD's lead Army public affairs proponency is likewise reviewing the career paths, training, and education for all its public affairs officers. For example, advanced degree opportunities are much broader, including such disciplines as mass communications, strategic communications, diplomacy, international relations, or even public administration. The Army recognizes its communications professionals need to be more broadly capable, culturally aware, and able to operate in volatile, uncertain, and stressful information environments.

The PAO is grounded in the operational Army through a base career as a soldier and a leader, commander and staff officer. Once entering the communications career field, this pentathlete can provide a broad range of communications capabilities to a commander. The PAO typically manages a portfolio that spans the full spectrum of information delivery, from internal product development to staff participation in the military decision-making process, to outreach innovation, legislative liaison, crisis communications, speech/testimony writing, communications operations, and strategic communications planning.

Army public affairs officers are already leaders, spokespersons, and Army champions, translators, and advocates. They are strategic communications planners, independent thinkers, and decision makers. Future plans are to broaden their experience base to ensure that PAOs are agile, flexible, culturally aware, sophisticated in emerging communications technologies, and savvy in dealing with all types of media. Additionally the notion of "broadening" career experiences for all Army officers is expanding—through the Joint, Interagency, Intergovernmental, Multinational (JIIM) opportunities program—while there are a number of other natural opportunities for an officer with this broad skill set to pursue: recruiting/marketing, legislative liaison, strategist, scholar, or interagency fellow.

Of late both the Army and the Air Force have placed individuals with operational backgrounds in the position of chief of service communications. Kenneth Bacon, a former reporter who became the Pentagon's spokesperson during the Clinton administration, has commented on this recent trend. "By far, the Navy and the Marines have been the most successful at public affairs," he said.[12]

In the Navy, in particular, he added, "they get these guys as young lieutenants, they work their way up through the system, and they know one of them is going to end up as Chief of Naval Information," the top Navy spokesman.[13] This is not true in the Army or the Air Force.

In his recent testimony before the House Armed Services Subcommittee on Terrorism, Unconventional Threats, and Capabilities, DASD (JC) Rear Adm. Frank Thorp agreed. "The Navy is the only military service to consistently promote Public Affairs professionals to flag rank," he stated. Now, "only one of the four services communication efforts are led by a career-qualified communication professional."[14]

So while the officers now heading Air Force public affairs have made a "good start," Bacon said, "if you really want to improve public affairs, you need to make it a productive career path: Build a strong cadre of young officers and promote them up the chain until one of them becomes the top person in public affairs."[15] The advent of broad-based strategic communications processes and the pentathlete concept for officer career development certainly makes this outcome possible for the Army's public affairs career professionals.

Vision

The emergence of strategic communications as a concept around which we can build solid, meaningful, and timely national communication of policy is logical and ripe for development. At the national level, our greatest asset is the recognition that from the seat of government, communications must be tied to national strategy and policy. Strategic communications is evolving as a process, one of necessity born in collaboration and integrated into every operation emanating from the National Security Strategy of the United States. Within the executive branch of government, we must be able to communicate consistently and clearly with America's allies and foes, as well as international audiences across the world stage, and remove the haze of suspicion born of mixed, changing, or incomplete messages.

In DoD our most promising efforts center on the evolving QDR Roadmap and ongoing efforts to organize, equip, train for, and support change in the communications field, while educating the force as to the broad range of capabilities this joint field can offer the joint commander. Strategic communications is not public affairs, but what it brings to public affairs is the strategic tie, focus, and structure.

In the Army the advent of strategic communications offers the resurrection of a small, historically marginalized career field, providing both challenge and opportunity for sophisticated career communications professionals. The door is open for these pentathletes to fulfill the need for strategic communications planning, to teach awareness and broaden the communications capabilities across the Army, and to provide strong communications support to the warfighter. This is the potential for strategic communications—to offer insight and understanding of how to apply information as a formidable element of national power.

Strategic communications is the process that serves as our route to the future, an acknowledgment of the need to craft communications with forethought, insight, and necessary ties to national strategy and U.S. government policy objectives. It is logically led by career public affairs officers who have the training, experience, capability, and potential to make it successful.

Toward Strategic Communication

*Communicating strategically during a war on global terrorism should be an
urgent part of the mission of every arm of the U.S. Government.
Explaining our government's actions and policies to the
peoples of the world must be a top priority.*[1]

• • •

A number of articles in the press in the past several years have reported that
political and military leaders are frustrated because the government does
not have an integrated process for delivering "strategic communication" on
issues of national importance, particularly the global war on terrorism. Frustration
over the inability to coordinate and synchronize public information activities has
been vented toward the Department of Defense and the military services. Others
have voiced similar concerns about a lack of cohesiveness and coordination within
the Department of State and the National Security Council (NSC). In short the
question of how to transform public communication channels and methods to
meet the challenges posed in an era of globalized, instantaneous, and ubiquitous
media has caused concern and even alarm. Moreover, many, especially in the
military, are worried that our enemies have already occupied and dominated the
information battle space.

Army doctrine has evolved greatly over the last three years to deal with this
challenge. It acknowledges that the information domain truly is a battle space and
that acquisition of favorable media coverage supporting regional and national
political objectives should be equated with seizing a form of key terrain. This view
is reflected, for example, in chapter 1 of the recently published FM 3-24, *Counterinsurgency*, which states, "The information environment is a critical dimension of
such internal wars and insurgents attempt to shape it to their advantage."[2] The FM
(field manual) clearly recognizes that counterinsurgent operations must be equally
sophisticated, flexible, and cognizant of the power of shaping information strategies.

Against such a background, let us ask, what is strategic communication?
How does it differ from the traditional means the government has used to inform

the public? For many, distinguishing between strategic communication and other, more familiar forms of public communication is either mysterious, problematic, or both. For some, determining what constitutes strategic communication calls to mind a comment by Adm. Ernest J. King, chief of naval operations, who reportedly said the following in the early days of World War II: "I don't know what the hell this 'logistics' is that Marshall [Army Chief of Staff Gen. George C. Marshall] is always talking about, but I want some of it!"[3] Many feel precisely the same about strategic communication. Although they do not know what strategic communication is, or how it works, they recognize that it is new and somehow more effective than older forms of public communication—and is therefore important. One result of this situation is that the expression "strategic communication" is one of the most misused and misunderstood terms in the military lexicon. The purpose of this chapter is to clarify the concept of strategic communications so that commanders at all levels can understand and exploit its benefits.

The Benefits: Why Strategic Communication Is Important

The principle benefit of strategic communication derives essentially from the same principle of war called "mass." Strategic communication means massing information among all agents of public information at a critical time and place to accomplish a specific objective. It avoids the destructive effects of mixed messages that result from not massing information. Dribbling out mixed, unsynchronized information instead of massing the release of unequivocal messages backed by a substantial body of facts is especially destructive during times of crisis or when the government and military find themselves under enormous public or political pressure, fastidious public scrutiny, and emotional criticism.

Many think the U.S. government habitually sends out mixed messages on issues of vital concern, messages in which policy is not clearly and consistently articulated or no clear justification for policy is provided. Such messages undermine confidence in U.S. policy by conveying the perception of disarray, vacillation, and weakness in the national will to any nation seeking to understand U.S. intentions. This frustrates allies, confuses potential friends, and encourages our enemies.

Our government's view concerning the Supreme Court ruling on tribunals is a case in point. The administration failed to provide a unified response to the court's ruling that military tribunals are illegal. Since the executive branch (including the Department of State, Department of Justice, and Department of Defense) could not or did not decide what unified message to promulgate regarding the ruling's significance to the war effort, widely different media interpretations abounded and went unchecked by a government public information counterweight. BBC News bluntly termed the ruling a "stunning rebuff to President Bush," and the French press generally followed a similar theme of "Supreme Court disavows Bush." German national radio hailed the ruling as a "victory for the rule of law." Civilian news media from Spain, Italy, Pakistan, and China agreed, while the Swedish newspaper *Sydsvenskan* editorial writer commented, "Now the judicial power has put a check on the executive power. Thanks for that."[4]

In contrast the Arab press reaction was skeptical. Writing in London's *Al-Hayat* Arabic-language newspaper, columnist Jihad al-Khazin commented, "This was all great news, so great that it was reported by all American and international media outlets and continues to draw reactions until this very day, but none of it is true, or, if we wish to be accurate, will ever see the light of day, because on the same day that the Bush administration declared its commitment to the Supreme Court's ruling, the Senate Judiciary Committee was holding hearings on the treatment of accused terrorists."[5]

How to Avoid Mixed Messages

We can describe the agenda-setting function of America's free press in the same terms the U.S. Army War College uses to define strategic leadership: telling people what to think *about* instead of telling them what to think. Strategic communication is an essential complementary activity to strategic leadership that manages public discourse not by attempting to tell people what to think, but by channeling information into the public information arena in an effective way. It sets the national agenda by establishing as a public priority what the public chooses to think about.

Strategic Communication Defined

To fully exploit strategic communication's potential to help people select what they think about, we must first distinguish it from other mere forms of public information and outreach programs. Doing so will help us define strategic communication, a necessary step to developing the rigorous training and education program leaders will need to enable them to focus on keeping issues of importance and the strategic messages concerning them prominently positioned in the national agenda.

Four major characteristics distinguish strategic communication from other types of public information:

Audience selection. Strategic communication differs from other public information activities in that greater care is exercised in the selection of audiences in order to achieve specific purposes. This stands in contrast to traditional public affairs and public diplomacy, the activities of which have been historically stovepiped—public affairs to U.S. domestic audiences, public diplomacy to foreign audiences. Moreover, most public information activities aim at broad public audiences. The Armed Forces Information Service, for example, targets the entire military community and the even broader general domestic audience interested in military affairs.

The first half of the definition of strategic communication set forth in the Quadrennial Defense Review's Strategic Communication Execution Roadmap particularly emphasizes the importance of audience selection: "Focused United States government processes and efforts to understand and engage key audiences to create, strengthen or preserve conditions favorable to advance national interests and policies through the use of coordinated information, themes, plans, programs and actions integrated with other elements of national power."[6]

Breaking down stovepipes. The QDR Roadmap definition also highlights the fact that strategic communication has a broader application than military public affairs. It calls attention to the need for formal mechanisms to compel a culture of cooperation among public information activities. In the past, public affairs, legislative affairs, outreach programs (academic, interest group, think tanks), and State Department public diplomacy essentially operated independently, within their own stovepipes, to reach different, discrete audiences. Consequently they sometimes addressed the same issues of public concern with contradictory messages and talking points.

The characteristic that distinguishes strategic communication from the old stovepiped way of doing business is formal cooperation among communicators. Strategic communications mandates that all public information agents in the government business—even coalition partners—must work together.

What distinguishes strategic communication from public information is a formal methodology that deconflicts messages through careful deliberation and coordination, analyzes and prioritizes key audiences, and synchronizes and times the release of information by all public information agents to their respective audiences in a disciplined fashion. Strategic communication also offers an opportunity to foster a true culture of engagement across the Army. In turn viable and active culture will drive and support the development of strategic communication as a force multiplier.

Public diplomacy in strategic communication. Under Secretary of State for Public Diplomacy and Public Affairs Karen Hughes has identified the objectives of closer coordination and integration among various government agencies dealing with public information and greater emphasis on developing cross-cultural capabilities. The State Department's public diplomacy effort is transforming the way the department does business.

Advocating increased funding for programs that are working, Hughes has mentioned international exchange programs, a direct form of community outreach (albeit on a global scale). She noted, "People who come here see America, make up their own minds about us and almost always go home with a different and much more positive view of our country."[7]

Another welcome change is that the State Department's emerging public communication strategy acknowledges the speed of global communications. The department has set up a new rapid response center based on the successful model used by Defense Public Affairs during the kinetic phases of the wars in Afghanistan and Iraq. The center monitors daily communications worldwide and provides a summary, along with America's response to diplomatic outposts. This information enables U.S. government representatives to be more knowledgeable and responsive U.S. policy advocates. The establishment of regional hubs to position spokespersons in key media centers such as Dubai ensures even greater presence and reach to key audiences in the Arab world. The department has also given senior regional representatives such as ambassadors and foreign service officers greater freedom to reach out to foreign audiences, both directly and through the civilian news media.

And, finally, the department has placed greater emphasis on using public diplomacy to shape policy. Noting that America hasn't always adjusted its programs to make their benefits clear to average people, Hughes said, the president "instructed us [the department] to look at ways to make programs more effective, to set clearer goals, focus our programs and partner with the private sector . . . then make sure we communicate what we are doing—a perfect example of the intersection of public diplomacy and policy."[8]

Rapid, comprehensive responses. The fourth element that distinguishes strategic communications from the traditional stovepiped operations that dominate much of the government's public information system is a rapid response that employs a range of communication tools in a synchronized, comprehensive way.

Strategic communication by its very name implies execution in support of a strategy, which in turn implies reaching specific strategic objectives. To compete in a global conflict in which lurid visual images and political messages often drive the agenda in compressed windows of opportunity, our strategic communication must be at least as efficient and speedy as our adversaries. To this end we are relearning daily that "being the firstest, with the mostest" in terms of initiative is just as applicable in the infosphere as on the battlefield.

Addressing this need in a speech in 2006 to the Council on Foreign Relations, Secretary of Defense Donald Rumsfeld commented on the Defense Department's view of the way ahead:

> *Government public affairs and public diplomacy efforts must reorient staffing, schedules and culture to engage the full range of media that are having such an impact today.*

Our U.S. Central Command, for example, has launched an online communications effort that includes electronic news updates and a links campaign, that has resulted in several hundred blogs receiving and publishing CENTCOM content.

The U.S. Government will have to develop the institutional capability to anticipate and act within the same news cycle. That will require instituting 24-hour press operation centers, elevating Internet operations and other channels of communications to the equal status of traditional 20th century press relations. It will result in much less reliance on the traditional print press, just as the publics of the U.S. and the world are relying less on newspapers as their principal source of information.

And it will require attracting more experts in these areas from the private sector to government service. . . . We need to consider the possibility of new organizations and programs that can serve a similarly valuable role in the War on Terror in this new century. . . . There is no guidebook—no road map—to tell our hard-working folks what to do to meet these new challenges."[9]

Defense Department efforts to improve public affairs to support the new imperatives of strategic communication began in 2004 during a "Tank Brief" on public affairs to the service chiefs of staff. That session was the result of a continuing debate centering on commanders' frustration with an ill-defined communications process. Following the brief DoD began to grow a strategic communication capability and structure, one supported by the findings of the QDR. Recognizing the importance of applying strategy to communication, DoD created the position of deputy assistant secretary of defense (Joint Communication) (DASD[JC]) in December 2005 to "shape DoD-wide processes, policy, doctrine, organization, and training of the primary communication-supporting capabilities of the Department. These include public affairs, defense support for public diplomacy, visual information, and information operations including psychological operations."[10] The terms of reference for the position state that it exists to maximize the Defense Department's capability to communicate in an aggressive, synchronized manner. The position clearly represents the first formal recognition of the need for a military communication advocate at the highest level.

One of the new DASD(JC)'s primary tasks was to improve all aspects of strategic communication by driving communications transformation in DoD and implementing decisions from the 2006 QDR. To this end a DASD(JC) working group developed a roadmap to provide strategic direction, objectives, milestones, and metrics for success. Just as important, the roadmap identified program and budget implications of strategic communication initiatives.[11]

The road map seeks to achieve three overarching objectives:

- Define roles and develop strategic communication doctrine for the primary communication-supporting capabilities: public affairs, information operations, military diplomacy, and defense support to public diplomacy.
- Resource, organize, train, and equip the DoD's primary communication support capabilities.
- Institutionalize a DoD process in which strategic communication is incorporated in the development of strategic policy, planning, and execution.

Furthermore, to address the fundamental requirement for strategic communication to be joint as well as interdepartmental and interagency, DoD initiated new requirements for joint public affairs officers. Despite a clear need for public affairs entities that have trained and worked together, there has never been a validated joint requirement for public affairs; consequently, there was no capacity. This omission laid the groundwork for failure in communication operations that developed rapidly and in an environment more global media stage and information oriented than anyone had anticipated—what commanders subsequently came to expect and want had not previously been explored or described in any detailed fashion. As a result the services were left to estimate, using their own doctrine, what they might need ad hoc. Given the situation, it should have been no surprise that capabilities did not match demands or expectations.

In addition to establishing the DASD(JC), DoD assigned formal responsibility for communication proponency by establishing a Joint Forces Command–based structure called the Joint Public Affairs Support Element. The JPASE exists to support the integration of communications into warfighter training; to develop operational public communication doctrine, program, and policies for the warfighter; and to give the combatant commander a rapidly deployable military public affairs capability at the beginning of an operation, when public communication is most critical and has the potential to be most effective.

The rapid and early deployment of a public affairs team in support of earthquake relief efforts in Pakistan was an early JPASE success. Within three days of the earthquake, the joint force commander had a team of operationally focused, culturally astute, trained professional communicators on the ground. Their presence gave the commander the ability to shape the information environment from the beginning of the operation, ensuring that actions and information fully supported the U.S. intent and goals. The team's ability to tell and amplify the global story of America's humanitarian efforts achieved the distinctly measurable effect of fostering greater understanding and more favorable views of the U.S. by international audiences.

Even as the QDR addressed the need to implement a culture of strategic communication within the Department of Defense via the strategic communications execution road map, the services were beginning to make sense of a broadly, but poorly defined and often little understood concept.

Working Together

The Army began to developing a strategic communication process in 2004 by establishing a strategic communications team within the Office of the Director of the Army Staff. The team's charter required it to link communications to Army strategy and priority programs, but it took nearly two years to mature the effort to the point where it could function as an integrated staff element and perform coordination and other tasks at the same level as the other divisions and branches. In April 2005 the responsibility for all Army strategic communication planning and the attendant staffing and funding for contract support transferred to the Office of the Chief of Public Affairs (OCPA). Using an enterprise approach to communications across the Army, OCPA began to develop strategic communication planning processes by building collaborative relationships with Headquarters, Department of the Army (HQDA) strategists, subject-matter experts, and other communicators. This created the structure, culture, and focus to support the development of Army strategic communication.

Another driving force was the Army senior leadership focus on developing a strategic communication capability. One of HQDA's objectives was to enhance strategic communication. With staff responsibility clearly in the Army Public Affairs portfolio, the objective was to "improve, over time, the strategic approach to Army communication, as well as the framework, mechanisms, customs, capabilities, and products needed for channeling the communicative energy of the entire Army." Army communication serves as the focal point for integrating "all Army efforts interfacing with a global public and should strive to be a 'best practices' benchmark for government, military and corporate communication." Everyone in OCPA involved in this effort has understood that "innovating communication within the Army Headquarters, and across the Army, demands a change in organization to create an enterprise approach to communication that better reflects the Army's current vision, mission, plan, and four overarching and interrelated strategies." The Army identified five lines of effort to drive this project forward: in process, structure, culture, image enhancement, and capabilities.[12]

The Strategic Communication Coordination Group moved to develop plans and associated products, such as the Army Communications Guide, which enhanced a variety of audiences' understanding of significant messages, campaigns, and events.

Senior staff support for applying strategy to communications and accepting collaborative planning processes in designing major communications campaigns is growing. Making public affairs the Army proponent for strategic communication is serving as a sense-making device, a construct that allows us to make sense of a new idea.

The Department of the Army has nested strategic communication planning and processes in the Army's strategy for transformation and solidly linked strategic communication to the national military strategy (fig. 4-1). This is significant. By beginning the hard, detailed, day-to-day work of establishing coordination and development/design processes for communications planning first at HQDA, and

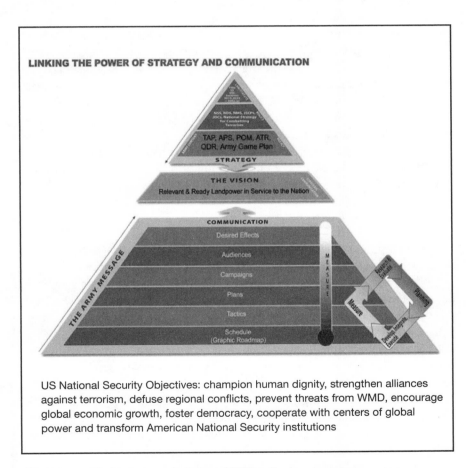

LINKING THE POWER OF STRATEGY AND COMMUNICATION

US National Security Objectives: champion human dignity, strengthen alliances against terrorism, defuse regional conflicts, prevent threats from WMD, encourage global economic growth, foster democracy, cooperate with centers of global power and transform American National Security institutions

FIGURE 4-1. Reprinted with permission of *Military Review* magazine.

in the next year throughout subordinate commands, the Army has taken the initial step to build an understanding of what strategic communication is and how strategic communication planning can work.

These efforts have already paid dividends in linking communications to the Army's long-term programs and process supporting transformation (fig. 4-2). As national concepts of strategic communication planning mature and DoD implementation of strategic communication processes evolve, the Army is ready to support and complement those efforts.

The Army is leading the effort to implement strategic communications throughout the DoD. The coordination group process and collaborative decision-making efforts that produced solid products have caused other organizations and activities to take a direct interest in the Army's progress. Members of OCPA's plans division have briefed the Army's process to the Office of the Secretary of Defense for Public Affairs, members of the Joint Staff, and the other services' communications leaders. While other activities may choose to adapt some or all of the Army's best practices, the Army is undoubtedly leading the DoD strategic communication effort forward. Even so, transforming to this new way of doing business across the Army will require significant, sustained investment in training and education at all levels in the future. Finally, for strategic communication to be successful, the Army must move the strategic communication concept of operations forward by fully resourcing the communications enterprise to support an expeditionary Army at war.

Building Strategic Communicators

More than twenty years ago, Maj. Gen. Patrick Brady, then the Army's chief of public affairs, said, "Clausewitz may not have listed information as a principle of war, but today it is, whether we like it or not. . . . There will be trouble if we ignore the need to inform our people and to deal with the commercial media in the planning, practice and execution of war. There is not enough training on public information in the military educational system. We are working on this issue."[13]

In some ways it appears that not much has changed since the day Brady made those remarks. Some individual tasks have been added to Army officer and NCO training courses; some courses remain unchanged. On the enlisted side, Army Basic Combat Training (BCT) first included a communications task in 2001. The Army Public Affairs Center first introduced this lesson to the BCT curriculum just after 9/11 and updated it in March 2005, when it added a lesson plan, "Interact with News Media."[14]

There is no public affairs or communications training in the Warrior Leader Course or in either the Basic or Advanced Noncommissioned Officer courses.

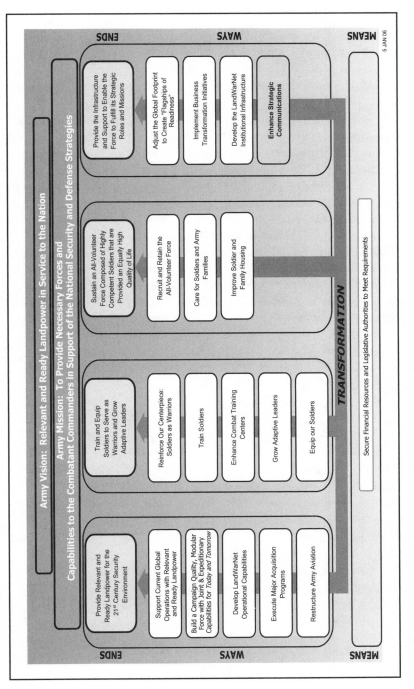

FIGURE 4-2. Reprinted with permission of *Military Review* magazine.

The first substantive training for NCOs occurs at the Sergeants Major Academy with a two-hour overview of Army Public Affairs. This is followed by a capstone command-post exercise in which senior NCOs participate in a media interview.

In the officer education system, the Basic Officer Leadership Course incorporates two hours of conference and discussion to train for successful participation in media interviews. A short practical exercise follows. U.S. Army Training and Doctrine Command (TRADOC) mandates two hours of media awareness training in all captain's career courses. This training focuses on company commanders and battalion staff officers supporting media operations in their area of responsibility. The Intermediate Level Education (ILE) course at Fort Leavenworth includes a two-hour overview of Army public affairs transformation to support current operations. Junior majors attend this year-long career course to prepare them for senior command and staff positions.

For years the Army's senior service college, the Army War College (AWC), held a "media day" for its resident students. The day consisted of panel discussions by members of the civilian news media, and officers were encouraged to bring their spouses. It was a day of grand entertainment conducted by media celebrities and did little to further any understanding or acceptance of a commander's responsibility to communicate or the necessity to plan for communications as a critical element of military operations. The Freedom Forum's 1995 report on the relationship between the media and the military, *America's Team: The Odd Couple*, scoffed at the educational value of these media days and recommended they be scrapped. In the past few years the AWC has, but it has not added any meaningful communications component to the core curriculum in their place.

The current core curriculum however, contains seminar discussions and exercises about the role of the media in the strategic environment. The AWC has also incorporated communications issues into multiple elective courses, and it exercises the students' abilities to conduct communications planning and media engagement in the course's capstone exercise.

Public affairs has always been closely identified with media relations because that is exactly what the Army teaches soldiers, NCOs, and officers in Army courses. But we teach nothing about internal communications, the importance of outreach in communicating with the American people, and the need for public affairs planning in operations; and of course, there is very little about the significance of applying strategy to communications or how to do it.

"The challenge is to train the force not what to think but how to think," Army colonel Peter Mansoor said in a 2007 interview with the *Boston Globe*. Mansoor, who led the Army and Marine Counterinsurgency Center at Fort Leavenworth, went on to say that troops must get inside the minds of the insurgents and the citizenry. "Counterinsurgency," he claimed, "is a thinking soldier's war. It is graduate-level stuff. There are public relations, civil affairs, information operations. It is not easy."[15]

Training and education, particularly in strategic communication, must be addressed across the force for strategic communication to succeed as an operational capability and for it to support DoD objectives in winning the battle of ideas. The Defense Information School is changing its curriculum to address the need for increased training in strategic communication, and Army Public Affairs has proposed the Senior Leader Development Office consider strategic communication training for colonels in its evolving professional development program.

The Public Affairs Officer

The aforementioned report on media-military relations, *America's Team*, was an extensive study that made detailed, exacting recommendations. It recognized the need for strategic public affairs leadership at the unified command level, stating, "In major conflicts such as Desert Storm, the Secretary of Defense and the Chairman of the Joint Chiefs of Staff should consider assigning an officer of flag or general rank in the combat theater to coordinate the news media aspects of the operation under the commander of U.S. military forces."[16]

This occurred at U.S. Central Command in the early days of Operation Iraqi Freedom. As operations in the Central Command theater began to generate operational velocity on the international stage, it became apparent the Public Affairs colonel did not have the staff muscle to serve the command at that required level. Rear Adm. Craig Quigley, a career public affairs officer, was then detailed from OSD Public Affairs to Central Command to serve as the director of public affairs. When Quigley retired, Jim Wilkinson, a former White House appointee with general/flag officer–commensurate rank, was assigned to take his place. When Wilkinson left at the conclusion of major ground combat operations, Central Command continued to look for a civilian of his stature, experience, and connections to take his place. That search was unsuccessful and the Central Command Public Affairs effort slowly began to revert to its prewar configuration and capability.

By the summer of 2004, U.S. Central Command's public affairs staff complexion had changed drastically, from a staff of seventy headed by a general officer or civilian equivalent to a staff of barely ten. The office remained functional despite having split operations between Tampa and Qatar; however, such a limited staff was unable to deal with the tempo of communications requirements, nationally and internationally, which had increased since the end of the conflict. This was not due to a lack of proficiency on the part of the staff, rather, it was a direct result of the immense, continuing demands of the global information environment.

Information Operations, as a communications capability, began to expand to fill that void, although later the overlap in mission sets was largely resolved with an expanded staff in the Public Affairs Office. Public Affairs generated a strategic communication approach to reaching American, allied, and Iraqi audiences and initiated an aggressive communications outreach focus.

Since then Central Command's public affairs operation has made significant strides, from responding rapidly to negative media pieces, to establishing a satellite office in Dubai's media city, to creating a team to monitor and respond to commentary in the blogosphere. Public Affairs professionals from all services have been responsible for tremendous innovation.

The Army public affairs officer is grounded in the operational Army by initial service as a soldier, leader, commander, and staff officer. Once entering the communications career field, this pentathlete can provide a broad range of communications capabilities to a commander. Public affairs officers typically manage portfolios that span the full spectrum of information delivery, from internal product development to staff participation in the military decision-making process, to outreach innovation, legislative liaison, crisis communications, speech-testimony writing, communications operations, and strategic communication planning.

Army Public Affairs officers are leaders, spokespersons, Army champions, cultural translators, force advocates, strategic communication planners, independent thinkers, and operational decision makers. Future plans are to broaden their experience base to ensure they are agile, flexible, culturally aware, sophisticated with emerging communications technologies, and savvy in dealing with all types of media. In addition, the notion of broadening career experiences for all Army officers is expanding—through the Joint, Interagency, Intergovernmental, Multinational opportunities program. An officer with this broad skill set can also pursue opportunities in recruiting, marketing, or as a legislative liaison, strategist, scholar, or interagency fellow.

The Army recognizes that its communications officers need to be more broadly capable, culturally aware, and able to operate in volatile, uncertain, and stressful information environments. Those who choose the public affairs field must understand this reality. Following DoD's lead, Army Public Affairs proponency is reviewing the career paths, training, and education for all its public affairs officers. For example, advanced degree opportunities are much broader. They now include such disciplines as mass communications, strategic communication, diplomacy, international relations, and public administration.

Vision

Strategic communications as a concept is logical and ripe for development. We can build a solid, meaningful, and responsive national capability to communicate policy around such a concept. At the national level, our greatest asset is the recognition that from the seat of government, we must tie communications to national strategy and policy. Strategic communication is evolving as a process. It was of necessity born in collaboration and integrated into every operation emanating from the National Security Strategy of the United States. Within the executive branch of government, we must be able to communicate consistently and clearly with America's allies and foes and with international audiences across the world stage. We must remove the haze of suspicion born of mixed, changing, or incomplete messages.

In DoD our most promising efforts are the evolving QDR Roadmap and ongoing efforts to organize, equip, train for, and support change in the communications field while educating the force about the broad range of capabilities this joint field can offer the joint commander. Strategic communications is not public affairs, but what it brings to public affairs is the strategic tie, focus, and structure.

In the Army the advent of strategic communication represents the resurrection of a small, historically marginalized career field providing both challenge and opportunity for sophisticated career communications professionals. The door is open for these pentathletes to fulfill the need for strategic communication planning, to teach awareness and broaden the communications capabilities across the Army, and to provide strong communications support to the warfighter at the strategic, operational, and tactical levels. This is the potential for strategic communications—to offer insight and understanding of how to apply information as a formidable element of national power.

The term "strategic communication" acknowledges the need to create communications with forethought, insight, and ties to national strategy and U.S. government policy objectives. It is logical that career public affairs officers who have the training, experience, capabilities, and potential to make it successful should lead it. Larry DiRita, former special assistant to the secretary of defense, said the headache of transformation is worth it: "The old-fashioned idea that you develop the policy and then pitch it over the transom to the communicator is over. You're continually thinking about communication throughout the course of the policy development process."[17]

Contrary to the view of some, strategic communication can be mastered operationally, its effectiveness can be measured, and it is distinctly different from other, more limited forms of public communication. However, the strategic communication process places a high priority on coordination and collaboration. This effort in regard to managing the release of public information has resulted in culture shock in both government and the media. The fallout has been many an emotional argument about whether such coordinated communication has converted government information provided as a public service into propaganda meant to manipulate not just our adversaries' perceptions, but our own peoples as well.

In a purely academic sense, providers of public information and purveyors of propaganda use similar if not identical communications tools (personal outreach, print media, electronic media, and computer communications). We must acknowledge that the government has a vital interest in political advocacy during a global conflict, and that globalization has changed to the rules of the public information dissemination. In an environment in which information travels instantaneously across national borders, when does simple prose aimed at providing public information become propaganda? Many question the legality of disseminating information to foreign audiences that clearly advocates on behalf of U.S. government policy positions when the same information ends up in American media channels, but such objections are unrealistic because all language inescapably both informs and influences.

The domestic media and other wary elements of the U.S. population fear that the coordinated use of powerful instruments of public communication and language will result in political domination through manipulation of the populace. This is not an unwarranted concern. Consequently the formulation of definitions that describe and differentiate types of communication, some of which could potentially be unethical, goes to the heart of the morals and ethics that underpin our constitution and democratic values—with direct implications for the information system the government uses to inform the U.S. public and the world. As the strategic communications process evolves and matures within the military and the U.S. government, such serious concerns will continue to surface. Unfortunately, there is no clear resolution in sight.

The Missing Element
Strategic Communication

Gatsby believed in the green light, the orgiastic future that year by year recedes before us. It eluded us then, but that's no matter—tomorrow we will run faster, stretch out our arms farther . . . and one fine morning—So we beat on, boats against the current, borne back ceaselessly into the past.[1]

♦ ♦ ♦

How are we to implement the strategic communication function in government? Since the twenty-first century began—what we now recognize as the war on terrorism, an era that truly started with the attacks of 9/11—our efforts in the United States have been limited. They have been restricted by our desire to right ourselves following those attacks and regain not only a sense of balance but also a sense of identity. They have been limited initially by what the 9/11 Commission report referred to so harshly as a "failure of the imagination." We failed to imagine that such an attack could or ever would be even possible or that our own aircraft could be used against us in such a way.

A larger failure of the imagination has occurred since 9/11. Our reactions have been defensive and not proactive or strategic—an ongoing failure to conceive a path ahead with purpose or power. We have launched wars and strengthened homeland security institutions but strategically have not imagined beyond a vague and naïve hope that the future will improve, and thus, as in the final words of *The Great Gatsby*, we falter and inevitably are brought back into the past, repeating the acts of before.

The issue goes beyond combating terrorism or even combating the ideological support for terror organizations. We must understand that our national identity is at stake and the American national response to acts of terrorism is insufficient to result in the prevention of future acts or deterrence. From containment to engagement, further to preemptive response and beyond, we must develop a national security strategy that recognizes the potential of strategic communication and resonates with the American people. We must develop an application of soft power that can last through the generational duration we now anticipate will be the nature of the war on terrorism. This newly forged identity must be one that can be generally supported and formed as an American enterprise—industry, government, and academia—approach to crafting a vision for the future.

Reaction, Not Action

The George C. Marshall European Center for Security Studies has sponsored a series of international conferences on how governments can counter both the ideological support for and appeal of terrorism. At the 2007 conference in Ankara, Turkey, participants from more than forty nations observed that gains thus far are for many Western governments primarily defensive in nature. They noted that those states, including the United States, tend to "address radicalization after it has occurred, instead of developing and implementing an effective preventive strategy that erodes the foundations of terrorists' motivations."[2] We react; we do not act.

Yet even as we discuss ideological deterrence, the notion of winning hearts and minds from the outside of a culture or ideology is doomed. We send messages internally; externally we project identity. It is a well-known tenet of public relations and communications planning that one way of building support and momentum for an organization's actions is to "preach to the choir." We have to cultivate the audiences who are listening to us, supporters, as well as those who may be indifferent, neutral, or undecided. It is a basic aspect of mass communications that one does not speak to those who are against our ideologies because they will not be listening anyway. We should continue to reinforce the message internally or find champions or secondary message carriers to carry it to other audience bases. Deterrence cannot work otherwise.

In a recent research study, a survey of more than six hundred journalists in thirteen Arab countries revealed that these journalists could be valuable allies and a conduit for explaining American policies to their audiences. That has not happened to date, and, the researchers noted, "America has failed to make use of what is potentially one of its most powerful weapons in the war of ideas against terrorism." Thirty-four percent of the journalists agreed, stating, "U.S. policy was the most popular answer to the question of what is the greatest threat facing the Arab world today."[3]

The war on terrorism is an ongoing, long-term crisis. At this point, as the horror of 9/11 fades in the American collective memory, our determination appears to falter, our commitment to fade. America has failed to make use of what is potentially one of its most powerful weapons in the war of ideas against terrorism. We are numb to the pervasive yet vague threat to our way of life, inured to the persistence of Code Orange billboards at airports and on interstates, and impatient with security searches of our person, removing water bottles, cosmetics, and shoes. And in the latter half of 2008, the lack of political will to move forward, from the seat of government in the Departments of State and Defense, the growing numbers of vacancies in key strategy, policy, and communications positions within the administration, and a stalled national security strategic focus leave our efforts to develop a strategic communication capability adrift, the backward movement inevitable.

A Travesty

In a recent Heritage Foundation paper, *Strategizing Strategic Communication,* authors Tony Blankley and Oliver Horn state frankly, "The fact that there is no national security strategy for strategic communication—or even a government-wide definition of 'strategic communication' seven years into the War on Terror is nothing less than a travesty."[4] Perhaps a greater travesty is the growth of boutique strategic communication organizations, in different parts of various Defense agencies, all with different missions and application of what each thinks its strategic communication turf is. Like weeds these organizations vary in size, structure, and focus, within DoD, State Department, and the combatant commands. Often there is a resulting duplication of effort and haphazard application of resources.

Our lack of a strategic communication strategy has provided a major advantage to our enemies. Al Qaeda and other terrorist organizations have effectively moved into the virtual battle space offered by the 24/7 global media environment and use the Internet and mass media tools to communicate with the citizens of its burgeoning virtual state worldwide. More than four thousand websites encourage the faithful, coax the uncertain, and effectively preach to that choir, unhampered by law, free press rules, or censorship. These groups have effectively wooed new members unopposed. And while operating freely in the vacuum, their communications have expanded in quantity, quality, and variety. Their internal messages are extremely effective; their external identity is becoming a recognizable brand.

As-Sahab, the media arm of Al Qaeda, released more than ninety videos in 2007, up from fifty-eight in 2006. In addition, the number of other distributors of violent recruitment videos and training academies is expanding.[5] Beyond messages of reassurance to the faithful, the content includes recruitment of suicide bombers and training academies, providing examples and information for would-be terrorists. How can this communications movement be openly countered?

Good guidelines and solid, significant recommendations abound. The Defense Science Board undertook the most comprehensive study in 2007. For this study the Strategic Communication Task Force members listened to a wide variety of sources in both government and industry on communication abroad. They heard from a number of successful enterprises where actions matter more than words.[6]

In their 2008 report, the task force provided a number of recommendations to the secretary of defense, and, more broadly, to the White House, to better position the U.S. government to conduct strategic communication in today's volatile security environment. In the report's introduction, Defense Science Board (DSB) chairman Dr. William Schneider Jr. noted, "The DSB first examined the matter of strategic communication in 2001, finding it an important instrument of national power."[7] The board provided significant recommendations then, in 2004, and again in this, the latest report.

The Way Forward

To undertake a transformation to build a strong network of strategic communication capability, the 2008 report recommended:

- Establishment of an independent Center for Global Engagement
- A permanent strategic communication structure within the White House in the form of a deputy national security advisor for strategic communication and advisor to the president
- Better use of existing structure and tools within DoD to support strategic communication
- Enhancement of the status, structure, and funding for the State Department's under secretary of state for public diplomacy and public affairs
- Review of the mission and capability of the Broadcasting Board of Governors as an integral element of this capability
- Creation in the DoD of a permanent deputy under secretary of defense for strategic communication, reporting to the under secretary of defense for policy[8]

Even without a detailed reading of the 131-page report, it is patently clear that the board addressed two major issues affecting the implementation of strategic communication as a function of government: first, the recognition and elevation of its function of strategic communication within the National Security Council and in the Departments of State and Defense, and second, providing significant funding for this function. Absent these two acts, further discussion of strategic communication as a valid application of the information element of national power remains an empty and purely academic discussion. Further, without this national commitment, any continuing bleatings in the press by either advocates or critics of the War on Terror as a protracted battle of ideas amounts to purely meaningless self-flagellation. We forfeit and the other side wins. Game over.

So is there movement? Is government making progress in branding the West, tarnishing the image and appeal of terror organizations, successfully empowering third parties to speak on our behalf, and continuing research into the dynamics of the global media environment?

One of the criticisms of government attempts to brand the West is that the act of branding or identity takes place continuously, and from the outside as well as from within. The U.S. government, whether engaged in public diplomacy or military action, is not the only entity actively branding the American nation and its values, goals, and ideals. Branding is also conducted as an independent process, a necessity one might argue, and it is done by a myriad of sources. Often the consequence is mere noise and dissonance, mixed messages, and confusion. It is rare that a coherent picture emerges. That which does emerge is often the result of even more than mass media analysis or reporting by the civilian press. It is the image projected by the American entertainment industry, one that in many cases cannot be an image of pride but, rather, one of decline, one that could be a true inspiration to terrorist goals.

A first step in understanding global media, in the words of Mark Maybury, is

> understanding how information flows among various media. For example, information posted on a web page might subsequently be reposted on a blog and invoke discussion, which might then be picked up as a story in a news site and eventually find its way to broadcast news. Understanding these often global flows is essential for modern public diplomacy.
>
> Just as Madison Avenue has developed methods (e.g., audience analysis, targeted marketing) to very successfully model and influence consumer behavior, so too governments need to skillfully leverage modern communications. Because media, and the information they convey, can be a force multiplier or a force divider.[9]

We Are What We Broadcast

The thought of MTV or even *CSI Miami* being viewed by young Saudis in Riyadh should be enough to make the average American swallow hard and resist the urge to cry out, "Our culture isn't like that. Really!" However, it remains evident that the shame evidenced by this export is telling enough. These do reflect our culture and national identity. It just is not a complete picture. Nevertheless the gratuitous sex and violence of average television shows and movies is certainly a reflection of a decadent culture where the veneer of civilization is wearing thin. Many times I have had to explain to European friends why it is safe to travel to the United States, that the country is not as violent as they have seen on television. I can recall clearly the day I stood in a classroom in the Swedish International Training Command several years ago trying to answer a question about Jerry Springer's reality show and why so-called average Americans wished to participate in it. It wasn't a program I had ever seen, and I was most uncomfortable in the realization that I represented that culture.

We have come a long way from the days of facing one enemy across the plains of Europe, nearly thirty years now removed from the 1980s. President Ronald Reagan was the "Great Communicator," the embodiment of American identity who stood tough against the Cold War threat. Even then some of America's greatest ambassadors were not elected leaders but, rather, representatives of American cultural excess, the mass media, and the entertainment industry. The characters from the television series *Dallas* were some of America's best-known citizens to those behind the Iron Curtain. Then, at the height of the Cold War, the wealthy, oil-drilling Ewings became the shining image of life in America, where it was easy to acquire money, cars, oil, land, power, and influence. Citizens from over one hundred countries, including many in the Warsaw Pact, watched the series faithfully, not with a disdain for the vulgar American capitalism as their communist leadership had hoped, but with a deep envy and yearning to acquire that lifestyle for themselves.

In April 2008 an article in the *Washington Post* argued tellingly that "*Dallas* won the Cold War." The authors state, "That lesson is more relevant than ever in an increasingly globalized world in which movies, music, and more cross borders with impunity—and the free West engages the semi-free East, whether in China or Iran. For all the talk of boycotts and bombs, the United States is interested in spreading American values and institutions, so a little TV-land may go a lot further than armored personnel carriers."[10]

The vast expanse of *Dallas'* Southfork ranch was one of the most compelling images America had in its arsenal of deterrence in the 1980s, and it was that cowboy image that conveyed so well the notion of security and prosperity. Strategic communication can serve as the strategic and policy power for developing an image of America in the war on terrorism, one that was not forged by accident but by design. That image, supported by existing culture and other message carriers, will be the one to carry the battle of ideas forward. Indeed, we must go further to create that image—and with fully conscious planning and awareness of the implications of what we create.

The recommendations of the Defense Science Board go a long way toward determining what this structure should be. A comprehensive and complete definition of strategic communication should likewise be considered. Before we can develop that strategy, however, there must be a solid government-wide definition. Attempts to date have been rendered so bland as to be nearly incomprehensible.

Finding a Definition

Jeffrey Jones, former director for strategic communication and information on the National Security Council, offered in a recent *Joint Forces Quarterly* article that strategic communication is the "synchronized coordination of statecraft, public affairs, public diplomacy, military information operations and other activities, reinforced by political, economic, military, and other actions, to advance U.S. foreign policy interests."[11] This is a full and comprehensive definition and completely illustrates why strategic communication will be ineffective if buried under an existing public affairs structure or within an information operations campaign. The function must be able to integrate actions, build consensus across stovepipes, and establish clarity of direction as well as unity of action. Otherwise, mixed messages result.

More alarming than the lack of unity is the emerging evidence of competition and discord. Each agency, combatant command, and organization is left to develop strategic communication capabilities as they feel best suits their mission, goals, and objectives. The Smith-Thornberry amendment (H.A. 5) to the 2009 Defense Authorization bill sought to remedy this through requiring a complete interagency strategy for strategic communication, specifying the roles and

Gen. David Petraeus responds to questions at a press conference at Iraq's Ministry of the Interior, 27 November 2007.
—*U.S. Army photo by Col. Rivers Johnson*

responsibilities within both Defense and State, and creating a Center for Strategic Communication.[12]

These provisions would have fully implemented the broad recommendations of the Defense Science Board. They would have moved the process of strategic communication in the American government to the point where it could successfully engage across the globe. The provisions sought to reestablish the American identity as that of the same people who stood behind the Marshall Plan and not that of those who advocate the abdication of moral responsibility. Unfortunately the amendment wasn't included in the final bill.

The American identity should seek to reaffirm what George C. Marshall referred to as America's position in his Marshall Plan speech of 1947: "An essential part of any successful action on the part of the United States is an understanding on the part of the people of America of the character of the problem and the remedies to be applied. Political passion and prejudice should have no part. With foresight, and a willingness on the part of our people to face up to the vast responsibility which history has clearly placed upon our country, the difficulties I have outlined can and will be overcome."[13]

It will take leadership and legislation, from the White House, Congress, and the Departments of State, Defense, and Homeland Security, to create a unified, comprehensive, and credible strategic communication capability. Understanding and harnessing this capability and its power to unify strategy and policy will move America's vision forward into the future and not permit it, like Gatsby, to be borne back into the past. But it must be developed. Soon, for the shore grows closer. And tomorrow is here.

The Golden Thread
Building Champions and Connections

A lie can get halfway around the world
before the truth can even get its boots on.
—*Mark Twain*

♦ ♦ ♦

In emergency medicine the "golden hour" refers to a time period lasting from a few minutes to several hours following traumatic injury being sustained, during which there is the highest likelihood that prompt medical treatment will prevent death.[1] The term has been in use in emergency medicine since World War I. While many trauma care experts may dispute the concept as a myth, it is well established that victims' chances of survival are greatest if they receive treatment within a short period of time following severe injury. Given the advances in emergency battlefield medicine in the past several years of the conflicts in Iraq and Afghanistan, the golden hour has sometimes been referred to as the "golden moment."

The concept of the golden hour translates roughly to a similar concept in today's news cycle. The golden hour of a major story, or even the golden moment at the onset of a topic's emergence, are critical to the life of the story and how it is perceived. This can also be understood in breaking news as the concept of whoever tells the story first and fastest leaves a lasting imprint on the public consciousness and gains control of the narrative.

In broad-based storytelling there are often multiple voices adding to the story as it builds. For this effort, although not necessarily collaborative in origin but certainly in result, a number of factors apply to shaping the narrative and its ability to "stick" in the audience's consciousness. Within communities, stories build membership and pride in shared experiences, and they can result in shared best practices. This is part of the golden thread that helps the story gain audiences, spread across cultures, and expand in resonance and understanding. That golden thread can be woven from many components, but there are a number of common factors in the process.

First the story must be authentic—speaking truth to power on a personal level (that is, it must be real and it must be instantly recognizable as real). This is critical as governments seek to engage internationally, breaking through outdated images, agenda-based politics, skewed reputations, and media-created perceptions: "Public diplomacy efforts go well beyond that of the U.S. Information Service (USIA), the Voice of America, and other media-driven approaches. An effective public diplomacy approach must include exchanges of ideas, peoples, and information through person-to-person educational and cultural exchanges, often referred to as citizen diplomacy. Years of successful exchanges have demonstrated the effectiveness of face-to-face interactions in breaking through stereotypes."[2] The golden thread can reach out and bring peoples together. Educational exchanges, social events, and shared experiences create new relationships and build trust. They make sense locally.

The story must be easily remembered and easily told. This concept of simple storytelling is at the heart of two popular business books, *The Tipping Point* and *Made to Stick: Why Some Ideas Survive and Others Die. Made to Stick* follows along the lines of *The Tipping Point.* The point is reached by social phenomena that cause ideas to tip from acceptance by small groups to large, whether in a viral manner or one pushed and prodded along by loosely formed networks of supporters. "Stickiness" is the appeal factor that enhances ideas' and stories' ability to spread and grow.[3]

The story must have a timeless and universal appeal. Its appeal is that which allows audience members to feel connected to the story. From there they develop their own stance, either to support and further the narrative or not. An example of this is the phenomenon whereby individuals in airports applaud service members returning from deployments. Whether they have been influenced by television commercials depicting the act, media coverage of large unit returns at holiday times, the request by an airline employee, or even the spontaneous recognition by other passengers, the result is the same. There is no question that thousands have experienced this social phenomenon in the past decade. It makes people feel good and it appeals to their sense of patriotism and community. They can choose whether or not to become part of the story.

Perception must align with reality. Images have a powerful role in building a narrative and in revealing its essential truth. The images from Abu Ghraib proved far more powerful than narrative descriptions of events in influencing the opinions of people, units, jurors, and ultimately governments and nonstate actors. As the Center for Stategic and International Studies notes, "The images of prisoner abuse from Abu Ghraib probably eroded America's moral authority as much as anything over the past six years because they seemed emblematic of a double standard."[4] Interestingly the media coverage could have taken a different turn in this case. Media criticism of the military was intense. However, the actions the Army took ultimately served to separate the values of the organization from the acts of individuals, and confidence in the institution itself remained high.

Weaving the Golden Thread

The golden thread may potentially be most influential at the onset of breaking news stories, but for communicators and champions alike, it is important to realize that the golden hour also can be thematically woven like a golden thread effectively throughout major information campaigns, programs, and even administrations.

The national fabric of our dialogue is frayed these days, and that fraying of the fabric by eroding confidence in the civilian news media opens it up to a greater opportunity for other voices and other viewpoints to not only be heard but also become strongly effective: "For decades, by deciding what stories were covered and how they were covered, newspapers set the boundaries of acceptable discourse in their communities. They reinforced normative community manners—and those communities allowed them to do so, with (for the most part) little complaint."[5] For many reasons that is no longer true, from the declining trust in mainstream media to the growth of the Internet and increasing transparency of information via online sources and the ability of many communities to link tightly on issues of social change, public policy, and other topics through new technologies and collaborative tools.

Organizations that seek to reach large audiences within specialized communities have a number of options available, whether a university with a large alumni population to reach or a chain of fast food eateries with a loyal customer base. In the military community, for example, there are a number of sites used by Army company-level officers to share insights, leadership challenges, and ideas. The company commander sites have been popular among junior officers. As the "digital natives" of a linked and connected age, they have their own community, and their connectivity builds upon shared personal and local experiences. Outsiders can observe and can contribute, but the greatest credibility comes from within. That is where the champions (experienced former company commanders) live and speak and influence one another and continue to build strong leaders.

I've personally used several systems in attempts to build community and empower champions, with varying levels of success. Within the Army Public Affairs community, I provided oversight to an official Army intranet site designed to provide access to guidance, policies, best practices, and successes in the communications enterprise across the Army. It was successful as a resource but less so as a means of generating dialogue and cross-talk. There is increasing dialogue across the career field now via Facebook and with the formation of an unofficial association, now a great point of pride among career Army public affairs officers.

I've personally used a "push" system of emails to the Army Reserve general officer corps to increase senior leader awareness of national-level issues and topics as well as inform them of the leadership's views on those issues. The push system, while highly effective as a personalized outreach tool, is still a one-way device, and although it generates responses, they tend to be responses directed to me and not shared. This is not a group of digital natives, and while many who grew up in the 1960s and 1970s can communicate electronically and do it well, it is still not second nature for them.

In providing oversight to the director and the staff coordinating support to the Army Reserve's corps of ambassadors,[6] I merged several of these communication methods and added and enhanced others. While communicating monthly via email to them all as a group, I also had to make accommodation for those ambassadors who did not use email—or even a computer for that matter. Similarly the website with information on policies and best practices received little attention initially. But when it was combined with a strategy of having each method of outreach and information market and support one another, the overall effect began to build.

Ambassadors were required to submit monthly reports of their activities, and I would highlight certain achievements in the messages I sent out monthly, encouraging best practices to be shared and promoting cross-talk. I particularly focused on those efforts several ambassadors supported in various communities, enabling them to both learn from and support one another. Many enjoyed hearing what others were doing in their communities, how employers supported deployed soldiers, and how ambassadors themselves could influence schools, groups, and civic organizations to provide support.

For example, for the Army Reserve's one hundredth anniversary in 2008, I asked all ambassadors to request from the governor of their respective states a recognition of that state's support for the Army Reserve. Many had never been to a statehouse or met their governor, and several were, at least initially, reluctant to undertake the effort. Yet as they began to learn of their peers' successes, they were encouraged, and they were pleased when they were able to achieve similar success. Naturally this opened the door to other types of relationship building, from discussions concerning issues with out-of-state tuition to veterans' support and job opportunities. It built as a grass-roots effort and that became its strength.

For the past several years, the Army Reserve's *Warrior Citizen* magazine has featured vignettes on ambassadors and their contributions. The fall 2010 issue featured Brig. Gen. Ashley Hall (Ret.) from Las Vegas, Nevada. He said, "Being an AR Ambassador provides me the opportunity. . . . To allow people throughout Nevada to give of themselves by becoming an Employer Supporter and providing financial

The tip of the iceberg: There is often more that lies just beneath the surface.
—*Courtesy of U.S. Army Public Affairs*

assistance and related assistance to help meet the needs and concerns of our Army Reserve soldiers and their families."[7] Hall is typical of these committed volunteers and gives his time to more than sixteen service organizations in Nevada.

The Role of Social Media: Weaving the Golden Thread

Even as some forms of connectivity have continued to flourish, others have proved to be less viable. "Blogging is no longer a primary way for people to express themselves online," and, according to eMarketer, outlets that provide a center for communities, such as Facebook and Twitter, have supplanted blogging. Yet more than half of Internet readers do read blogs, and that number is projected to rise even higher in the next several years. The idea that blogs have become "an accepted part of the online media landscape" is one reason for the readership increase.[8] Again, widespread acceptance of technology is what counts. Even as the twenty-four-hour news cycle began to claim precedence in the early 1990s, and the cellular telephone became ever smaller and more portable, there were outcries against the speed of communication and the increasing pressure for response and commentary. But the resulting community and connection was the greatest benefit of these advances— that and the democratization of the communications process.

Another plausible reason is the ongoing building of communities based on interests, views, and other aspects of human interaction other than geography. While thinking globally and acting locally, many feel more connected to others with like ideas, thoughts, or interests. Large numbers of service members, both those who are deployed and who have returned, now follow more than the news media coverage of the conflicts in Iraq and Afghanistan. With even more keen interest they follow and contribute to the blog posts of their peers who are deployed in commands and outposts from Kabul to Baghdad.

When one spokesman commented on his Facebook page that he had spent about two hours with a well-known radio outlet on an extremely contentious story only to later discover that the military's viewpoint received less than thirty seconds of airplay in a four-minute piece, many other public affairs officers posted on his "wall" their understanding, support, and empathetic outrage at the treatment he'd received.

The Golden Thread in Niche Markets

Certainly the reported death of mainstream media may be premature or at least overrated. While the issue of distrust in media reporting remains high, there is still an audience and there is in fact greater interest and involvement in the news process. According to the Pew Research Center, the "average time Americans spend with the news on a given day is as high as it was in the mid-1990s, when audiences for traditional news sources were much larger." Traditional outlets, particularly magazines, continue to decline slowly, but digital platforms more than make up for that decline. About 36 percent of Americans, according to a Pew poll in September

2010, stated they get their news from both traditional and online sources.[9] This mix will undoubtedly continue over the next several years, although the percentages may differ. We now have both the horse and the brand-new automobile in the barn, and they are vying for space, resources, and our attention. The point is that the advent of social media has not immediately destroyed traditional media, as many feared. All of these forms have become mutually reinforcing.

Many advertisers are willing to point out this relationship as well. As newspapers tout new iPhone apps, television and radio news direct their viewers and listeners to related websites. Magazines are pleased to remain popular. A recent Ad Council campaign, appearing in magazines from fashion to travel, compared the relationship of magazines to the Internet to that of coffee with instant coffee:

> *New technologies change many things. But not everything. You may surf, search, shop and blog online, but you still read magazines. And you're far from alone. Readership has actually increased over the past five years. Even the 18-to-34 segment continues to grow and typical young adults now read more issues per month than their parents. Rather than being displaced by "instant" media, it would seem that magazines are the ideal complement.*
>
> *The explanation, while sometimes drowned out by the Internet drumbeat, is fairly obvious. Magazines do what the Internet doesn't. Neither obsessed with immediacy nor trapped by the daily news cycle, magazines promote deeper connections. They create relationships. They engage us in ways distinct from digital media.*
>
> *In fact, the immersive power of magazines even extends to the advertising. Magazines remain the number one medium for driving purchase consideration and intent. And that's essential in every product category.*
>
> *Including coffee.[10]*

The issue of building community via community champions who advance ideas, policies, and notions toward a general tipping point is not only at the heart of how community outreach programs work but, beyond that, constitutes a personal form of engagement. They can truly form the heart of what is a genuine "soft power" strategy, through relationship building.

Champions provide that golden thread, the immediate up-close and personal representation of an administration's viewpoint, a corporation's product, or a public figure's stance. Those representatives, loosely tied to a central point by current and sought-after information, are the true movers, ones who can build upon new ideas, increase their acceptance, tell stories illustrating success, and build toward a tipping point of broad acceptance and understanding. Little things, then, can make a big difference.

It is networks and communities that build consensus, networks that influence product acceptance, publicly dismiss wrongdoing, or refute often-spurious claims by individuals seeking to defraud or negatively influence. Reporters may claim that the role of watchdog is and has long been rightfully theirs, that it is the reporter who has the primary seat as a witness to history, whether on the battlefield or at a school board meeting. But social media has leveled that playing field and opened the door to a loud and often raucous cacophony of voices. Facebook, MySpace, and others have a definite place for the citizen as witness and participant. Viral videos on YouTube are likewise examples of the influence that average people can generate. These disparate voices clamor for their equal and legitimate right to be heard.

They may serve not only as witness or town crier but also as the debunker of lies: "The great enemy of truth is very often not the lie—deliberate, contrived, and dishonest—but the myth—persistent, persuasive, and unrealistic."[11] This, then, is the role of the golden thread, to serve as the great equalizer. Mark Twain noted the speed with which a lie can travel around the world; it moves even faster when covered with a glistening coat of salacious gossip or intrigue. Conditioned by years of television viewing and exposure to movies in which conspiracy theories reign supreme, we need the grounding of simple facts, whether delivered by friends who text from the scene of an event or from champions we can trust with the truth. They have credibility stronger than pervasive myths and can prove them false.

The Toxic Information Environment

I found it astonishing that one person can actually have as big of a voice online as what an entire media company can on Twitter.[1]

• • •

Government application of "soft power" via public diplomacy, communication, and building partnership capacity is necessarily dependent on making solid connections. Soft power is realized through viable, strong communication networks. It is built through action but is supported with messages that are consistent with national identity and have the ability to influence as long as they align with the power of truth.

While we have long concerned ourselves with the finality of our definitions as to what constitutes information operations versus public affairs, psychological operations (psyops), or public diplomacy, the atmosphere in which we communicate—internationally, nationally, locally—continues to grow ever more toxic. We are at the point now where definitions matter little, and by consensus we must realize that all communication, even that which merely professes to merely inform and educate (not to mention entertain), also influences.

The communication atmosphere has continued to fragment and deteriorate since 1994, from the day O. J. Simpson stepped into his white Bronco and turned the key to the moment Tiger Woods sent his first text message to a potential new girlfriend. Broadcast news transformed from its early days broadcasting community-relevant events, issues, and serious topics to a medium focused on entertainment and titillation. It grew in scope beyond the coverage of immediate events of crisis and calamity to regularized "seasons" of events coverage, such as those of coronation (public ceremonies ranging from inaugurations to papal elections, royal weddings, and elections), contest (sports, politics, and business), and conquest (war coverage, diplomacy, and unrest). We have become in our news appetites like junk food addicts, obese and complacent with our daily servings of hydrogenated mindless fat and sugar.

Spurred on by a weakened economy, we have arrived at the critical decision point for mainstream media to either choose life, that is, ethical and relevant coverage, plus a renewed sense of purpose in community service with a generous helping of editorial rigor, or a continued and ever more rapid slide toward irrelevance and extinction. Communication in our current age, that of cynicism and celebrity, wavers between the dangerous and the absurd at all levels.

Coronation

News has often been called "the first rough draft of history." In the mid-1990s there was still a majority reporting in this news category, despite the beginnings of an all-encompassing fascination with celebrity and the growing public cynicism with politics and news coverage. Sensation then was still a condiment, not the main course it has become today. "In 1995," author Sam Tanenhaus notes, "cable news remained the bland civic pasture of CNN and C-SPAN; Fox News and MSNBC were not founded until the following year."[2]

Journalism schools still proudly touted what they termed the agenda-setting function of the civilian news media, not to tell people what to think but to tell them what to think about. This is the function of those who report the news in a democracy, to report what happens without bias, to serve the community with what its citizens not only want but also need to know. For small town newspapers and television stations, this has meant covering town hall meetings, mayoral and city council elections, the boring but essential mechanics of a school board session, high school graduations, social and community events, and all the other elements of community life, from births to weddings to deaths. On a global scale, this included national elections, international diplomacy, and the coronations of kings and popes. Audiences watched and they participated.[3]

Over time coronations began to acquire a more salacious note. Long before the wedding of Princess Diana and Prince Charles, the unblinking media eye focused ever more sharply on the everyday lives of leaders, heroes, and saints, and that naturally brought their failings into the spotlight. As mass media ownership evolved to media giants and company groups, even smaller outlets began to focus on news that was exciting and capable of generating controversy and interest, not to mention subscribers and viewers. The mantra "If it bleeds, it leads" became ever more apparent as local newscasts focused increasingly on stories of crisis, calamity, and crime. Polls still show this phenomenon and its influence on local coverage and politics. This likewise follows the negative coverage trend at the national level.

Major Trends

A recent *New Yorker* magazine article on media culture and politics focused on a prescient comment by Richard Hofstadter. In 1954 he said, "The growth of mass media of communication and their use in politics have brought politics closer to the people than ever before and have made politics into a form of entertainment."[4] Witness the number of politicians who write wildly self-aggrandizing and self-justifying "tell all" books about their own careers in government service, personal missteps, and great accomplishments. Failures are often glossed over or just omitted.

This prediction was borne out by the Project for Excellence's annual "State of the News Media" report in 2010. The report noted that the future of both new and old media are more tied together than may be apparent. For example, as online numbers continue to grow, many online news sources rely on what is reported in mainstream media for initial sources. In a domino effect, as mainstream media continue to experience cutbacks, the public receives less information. Indeed power appears to be shifting to the newsmakers since they have the ability to provide the first accounting of events and policy changes.[5]

Contest

Even reality television shows, which began with shows like *Cops* and other police docudramas in the mid-1990s, were more like documentaries than contests between "the good guys" and "the bad guys." However, the concept of "contest" as a format was, although not noble, at least governed by propriety and clear, civilized rules. As contests evolved into "survival" at all costs, they contributed mightily to the decline of polite discourse and the strained veneer of civilization became ever more transparent.

Sportsmanlike conduct also began to rapidly unravel at this time. Professional sports contests were plagued by players and owners with gambling and other addictions—as well as criminal arrests and behavior, both on and off the fields of play—that would make any traditional role model cringe. Ask Pete Rose, Mike Tyson, Mark McGwire, Darryl Strawberry, or any of a dozen other flawed professional athletes over the past decade who watched their careers and reputations publicly go up in flames.

Reality television programming became less real as it abandoned the limitations of plain, straightforward documentary and began to morph into vicious contests of elimination. The reward was the standard fifteen minutes of fame on television in front of massive voting audiences, some larger than previous votes cast in presidential elections. While talk shows featuring dysfunctional family dynasties and staged violence faded, the reality format continued to evolve. Game shows still existed in the margins, although the 1990s British show *The Weakest Link* came to epitomize the dumbing down of the genre with its caustic farewell lines: "You are the weakest link. Goodbye."

The contests flourished. This genre's popularity grew steadily from the first shows featuring bachelors, island survivors, and business apprentices to shows featuring drug addiction interventions and competitions in cooking, fashion design, bisexual dating, homosexual home decor, and boxing.

Conquest

Stories about conflict, diplomacy, and international relations are still relevant and serious. War and conquest is one of the oldest subjects in history, recalled and recounted, but today those stories embody not only the best elements of other types of communications coverage but also the worst. The military is not immune to coverage that leans to the celebrity—although coverage of military heroism is statistically much less relevant than coverage of celebrity. Witness the story of former pro football player Pat Tillman, who joined the Army after 9/11 only to die in Afghanistan as the result of friendly fire. Consider the lingering doubts left by the news media as to the nature of his death. Coverage of military operations is likewise fraught with so-called analysis, with the prevailing tone one of skepticism, if not outright disbelief concerning the commander's motives or efforts.

Perhaps one of the greatest risks in military information operations today is that the media covering events and analyzing operations are less experienced than ever. Downsizing national news organizations, shrinking news magazines, the rise of the blogger, and a weakened economy have seen a wholesale exodus of talent and experience for the door. There are legions of younger, less experienced reporters now on the military beat, but fewer of them overall. Smaller numbers of reporters are chasing larger stories, and the pressure to be first is significantly greater than the pressure to be factually correct. News has shifted from putting things into perspective to blowing them out of proportion.

The Internet continues to evolve and mature. To some the information domain may look like an undisciplined, wild landscape. Others view it as a fledgling democracy in its own right, where news and information and eyewitness accounts and opinions all exist simultaneously with the most accurate, factual, verifiable, and logical sites gaining in viewers and reputation. The immediacy the information domain provides is a major threat to the traditional news media. The crumbling credibility, loss of a sense of community hastened by a decade of exposes, bad reporting, lies, and lawsuits only appear to be hastening the downhill spiral and increasing irrelevancy of much of the mainstream media. Many appear to have lost their own identity as they struggle to identify means of financial survival and try to ascertain how they should partner with new media or use it to their own advantage.

The documentary, however, is one news category that has actually expanded in the past ten years. While it provides more in-depth, extensive coverage of a topic, it also requires more resources to support. Often this can be offset by the opportunity to continue to garner new audiences through repeated showings and can continue to gain in influence. Certainly Al Gore's documentary on global warming is a prime example of a documentary with a huge mainstream audience. The docudrama produced by Home Box Office several years ago, *Baghdad ER*, is a prime military example of a controversial documentary subject that many Army leaders found uncomfortable, but it was well received by the American public. Since the global war on terror began in 2001, HBO has produced more than thirty Iraq War–related documentaries, including such well-known programs as *Alive Day: Home from Iraq*, which tells the stories of returning wounded soldiers, and *Wartorn: 1861–2001*, with its focus on post-traumatic stress and suicide.

Documentaries, like many of the antiwar HBO programs, can be agenda based and have a much longer shelf life and over time attract larger audiences than a single news story that only airs once. Movies such as *Stop Loss* and *The Messenger* likewise bring Hollywood's perspective to military stories, while Kevin Bacon's moving portrayal of a Marine lieutenant colonel in the HBO movie *Taking Chance* focused on the community aspects of the officer's journey as he escorted a young Marine's body to his hometown following his death in combat. Entertainment also influences opinion.

One additional factor to consider is humor. Jon Stewart's *Daily Show*, with its scathing commentary and often-critical view of the mainstream media, has become a major news source itself in the past several years. Stewart, in fact, was named to an American top ten celebrity list: "People Who Change the Way We Think in 2009." Apparently society's thought leaders are primarily entertainers. NBC News anchor Brian Williams commented, "Jon Stewart has created a system of checks and balances. On occasion, when we've been on the cusp of doing something completely inane on *NBC Nightly News*, I will gently suggest to my colleagues that we simply courier the tape over to Jon's office, to spare *The Daily Show* interns the time and trouble of logging our broadcast that night. That usually gets us to rethink the segment we were planning on airing. How did we ever live without his show's sharp tongue?"[6] This is certainly a telling statement and one indicative of the naïveté of a profession now faced with criticism for the first time, a profession that long refused to discipline itself.

One might as well ask what we did before we became so cynical:

> Culturally, this has been the decade of the reality show. And what do we have to show for it? Survey the wreckage. The marriage of the octoparents, Jon and Kate, is a shambles. Richard and Mayumi

Heene were so desperate to land a reality series that they concocted an enormous hoax, convincing the country their child had been carried away in a balloon. Michaele and Tareq Salahi tried to claw their way onto the sure-to-be-hideous series Real Housewives of D.C. *by brazenly sashaying into a state dinner at the White House. The British historian Arnold Toynbee argued that civilizations thrive when the lower classes aspire to be like the upper classes, and they decay when the upper classes try to be like the lower classes. Looked at through this prism, it's hard not to see America in a prolonged period of decay.*[7]

Challenges and Dangers

The challenge of communicating in this toxic environment is not only one of cutting through the pollution but also one of getting through the clutter to a market that is already saturated with millions of messages designed to sell and influence. Government and military communications can remain relevant and effective in this environment through strict adherence to immutable ethics and through acknowledgment that in a saturated, polluted communication environment, words mean little. Words may mean little but they must be transparent. Actions matter and they must be consistent. Relationships last.

The annual Harris poll for 2010 posed the question, "How much confidence do you have in the leadership of the following sectors?" (see fig. 7-1). Not surprisingly, the military was in first place, with the medical community and nonprofits in second and third place respectively. Wall Street was last, with the press not far behind. Broadcast news rated slightly higher.[8]

While mass media sources have grown more global, studies show that our references are becoming ever smaller. We tend to trust the information we receive inside our own communities, that which we hear from friends and colleagues, people who share the same values and have the same interests in the same subjects. We believe what we know to be true within our own communities rather than how we see it described from without. What the Harris poll and similar surveys tell us is that Wall Street and the mainstream news media are disconnected from our communities. They do not represent our interests and we do not trust them.

One company has made a popular practice of sharply playing to our beliefs and our cynicism. At Despair.com, there are numerous workplace posters and t-shirts for sale that ridicule business beliefs and values and criticize the media. One poster depicts a picture of a television in a living room, on with a blank screen, before a sofa. The caption reads in large print "PROPAGANDA: What lies behind us and lies before us are small matters compared to what lies right to our faces."[9]

CURRENT CONFIDENCE IN LEADERS OF INSTITUTIONS (2010)

"As far as people in charge of running (READ EACH ITEM) are concerned, would you say you have a great dal of confidence, only some confidence, or hardly any confidence at all in them?"

Base: All Adults

	A Great Deal of Confidence	Only some Confidence	Hardly Any Confidence At All	Not Sure/ Decline to Answer
	%	%	%	%
The military	59	30	9	2
Small business	50	42	5	3
Major educational institutions, such as colleges and universities	35	49	13	3
Medicine	34	47	16	3
The U.S. Supreme Court	31	46	21	2
The White House	27	38	33	2
Organized religion	26	44	24	6
The courts and the justice system	24	54	19	3
Public schools	22	54	22	1
Television news	17	54	26	3
Major companies	15	56	27	3
Organized labor	14	49	31	6
The press	13	47	39	2
Law firms	13	54	28	6
Congress	8	41	48	2
Wall Street	8	43	45	4

Note: Percentages may not add up to 100% due to rounding.

FIGURE 7-1. Results from the Harris annual poll, "Current Confidence in Leaders of Institutions (2010)."

The 2.0 Community

Pervasive cynicism leads to prevailing mistrust. The changes in coverage in the categories of conquest, contest, and coronation coupled with increasing cynicism and public displays of vulgarity in the first decade of the twenty-first century have nearly resulted in communications gridlock. We automatically distrust the mainstream news we receive, the explanations from leadership, and the explanations we see daily. With the Internet as the backdrop to new developing social norms, there are several factors that influence this sense of community:

1. Enormity—a much larger scale network, larger numbers
2. Communality—a broadening scale
3. Specificity—increase in specific ties
4. Virtuality

Many people are familiar with the movie *Six Degrees of Separation*, but in the 1960s, when Stanley Milgram experimented with the phenomenon of separation, he went one step further than six degrees, adding what he termed the "three degrees of influence." He believed that everything ripples "through our network" and impacts our friends, our friends' friends, and even our friends' friends' friends. Community influence tends to dissipate thereafter. He posited that this applies to a wide range of attitudes, feelings, and behaviors.[10]

These degrees give a measure of safety and security in an increasingly toxic communication environment. A sense of balance and trust comes from within secure groups and firsthand knowledge or experience. What can the mainstream media offer in this interactive atmosphere? The answer may lie in those aspects of national news coverage missing now—balance, objectivity, ethics, and perhaps the ability to recognize and cover trends.

The mainstream media certainly didn't see the collapse of the economy coming, despite warning signs. Since the media, particularly print media, cannot compete with social media's speed or personal knowledge's trust, the one area where it can provide a unique service and real relevance is in the realm of positing and preparing for the future.

The challenge for military leaders in information operations is rendered exponentially more complex in this toxic and volatile environment. The internal push and the internal pull to protect information versus releasing it are very real. Witness the issue of social media access via government computers explode over the past several years and the now reluctant opportunity for access and mixed results. Perhaps we sometimes ask the wrong questions. Rather than ban social media access on military computers, we should focus our efforts on ensuring that classified systems do not have parts or capabilities that are so interchangeable with nonclassified systems. Today leakage is not only a possibility but also extremely difficult to prevent. This is a true design flaw (problem) and not a user issue (symptom).

The gates of a military base can't keep the toxic atmosphere at bay. Soldiers, sailors, airmen, and Marines read *People* magazine, vote for their favorite singers on *American Idol*, and laugh at Jon Stewart's antics on *The Daily Show*. The toxic atmosphere is pervasive, and we drink the Kool-Aid because so often it is all there is to drink. There is no artificial way to dismiss it or the blogs service members create themselves, the communities they form, and the actions they take, whether or not they are in concert with this institution that is so highly respected in America.

Coronation, Conquest, and Contest

"The time has come for the world to move in a new direction. We must embrace a new era of engagement based on mutual interest and mutual respect, and our work must begin now."[11] As President Obama spoke these words before the United Nations General Assembly last fall, the information battle continued to rage quietly in cyberspace. Operational security efforts continued to educate service members and strived to protect information from falling into enemy hands. Military deception planners struggled to devise programs to deceive and delay adversaries in their quest for a competitive advantage in the information environment. Information engagement efforts continued to progress, although they were sometimes disconnected from other efforts and sometimes proceeded blindly. Nevertheless all lines of operation forged ahead through the noise and the pollution.

The atmosphere is indeed polluted. But the respect our military has, both at home and around the world, ensures that we will continue to be a global leader in outreach, engagement, and nation building. The ongoing information battle in cyberspace, as well as the more visible effort to win public support, ensures that we will continue to forge ahead in building communities of common interest, friendships and business relationships, and shared understandings on global issues such as the environment and use of natural resources. In the long term this is what can truly transform culture and build civilization.

The Cowboy and the Soldier
Military Communication Ethics

Courage is being scared to death—but saddling up anyway.
—*John Wayne*

◆ ◆ ◆

The cowboy has long symbolized the ideals of the American dream. The American dream is one of freedom, independence, resilience, and simplicity. It is a hero's life as lived by the quintessential American knight, one with a broad hat, a horse, and a gun. The cowboy is an adventurer in hundreds of folk stories; he faces many seemingly insurmountable challenges but always prevails. He has been the star of stage and screen, fighting bad guys, saving the day, getting the girl, and always embodying the highest of standards.

That twentieth-century hero, the American cowboy, has in the twenty-first century transformed, in both heroic image and reputation, into today's most visible patriot and role model, the American soldier. The relationship between the two is easy to discern; both have an indelible emphasis on basic values. Essentially, cowboy ideals are the same values expressed in the Army's "Warrior Ethos." The ethos goes to the soldier's identity and can be found in a number of posters, articles, and discussions on the Army's website (www.Army.mil):

> *I am an American Soldier.*
> *I am a Warrior and a member of a team.*
> *I will always put the mission first.*
> *I will never accept defeat.*
> *I will never quit.*
> *I will never leave a fallen comrade.*
> *I am disciplined, physically and mentally tough, trained and*
>   *proficient in my warrior tasks and drills. I always maintain*
>   *my arms, my equipment and myself.*
> *I am an expert and I am a professional.*

I stand ready to deploy, engage, and destroy the enemies of the
*   United States in close combat.*
I am a guardian of freedom and the American way of life.
I am an American Soldier.

The ethos was brought to the U.S. Army by its thirty-fifth chief of staff, Gen. Peter Schoomaker, who grew up in Wyoming. At his farewell ceremony in 2007, Pete Geren, then acting Secretary of the Army, said, "Pete Schoomaker's list of accomplishments is long and will prove enduring, but perhaps the one that has done more to prepare our Army for whatever lies ahead is the Warrior Ethos. The words he requires every Soldier to learn by heart and an ethos he has lived every day he has worn the uniform. Every Soldier must learn it, but they must more importantly learn what it takes to live it. No matter the rank, the job, the unit, all Soldiers are Warriors. They learn to always place the mission first; never accept defeat; never quit; and never leave a fallen comrade. The ethos is and will continue to be the soul of the American Soldier." Schoomaker approved the Warrior Ethos on 24 November 2003, and it was first published in *Infantry* magazine in December 2003.

Geren continued: "Pete has also said, 'If you're riding ahead of a herd take a look back every now and then to make sure the herd is still there.' Pete, you can be assured that this herd is following you every step of the way and that your leadership will be missed."[1] General Schoomaker was often fond of quoting pithy cowboy commonsense stories. He effectively merged the cowboy's tough, commonsense approach to life with the soldier's ethos of service to country.

In 2004, at about the time the warrior ethos was gaining broad acceptance across the Army, Wall Street veteran James P. Owen published a book titled *Cowboy Ethics: What Wall Street Can Learn from the Code of the West*. Owen spent thirty-five years in the investment management industry. He notes somewhat wryly that it is not an industry that often gives practitioners an opportunity to do something from the heart, which is exactly what he attempted with this book. In the dark days after Enron and a host of other scandals, large and small, Owen found himself thinking that only a return to basic values could help the troubled industry find its way again. His inspiration was based on the life of the working cowboy and his code of ethics. The picture book *Cowboy Ethics* was the result.[2] The code discussed in the book closely parallels the Army's Warrior Ethos.

The values of the cowboy and the soldier are much alike: honor (adherence to the Army's publicly declared code of values), integrity (possesses high personal moral standards, honest in word and deed), courage (manifests physical and moral bravery), loyalty (bears true faith and allegiance to the U.S. Constitution, the Army, the unit, and the soldier), respect (promotes dignity, consideration, fairness, and equal opportunity), selfless service (places Army priorities before self), and duty (fulfills professional, legal, and moral obligations).[3]

U.S. Army values poster.

More than twenty-five years ago, author Ben Stein called on the cowboy as a reference for his views on commonsense problem solving in business, or what he termed "bunkhouse logic." This is the cowboy's way of reasoning, Stein argued, to overcome obstacles and to complete the mission. Stein explained that bunkhouse logic requires constant activity and an emphasis on performance, not excuses.

Interestingly he noted the applicability to business ethics, much as Owen would later: "By an extraordinary chance, the skills I observed at ranches . . . are amazingly similar to the attributes I detected in the managing director of an investment bank and the CEO of Dow Jones."[4]

Stein went on to apply bunkhouse logic to all undertakings in career and life. It is the mindset that makes the difference, the acceptance of risk as inevitable, healthy, and desirable, a mindset that maintains a steady focus on forward movement. This pursuit of success and happiness comprises the quintessential American folk story.

Folk stories resonate with Americans. These are stories that are recognizable as intricate to the fabric of the American life; they are deeply rooted in the American culture and experience. They are Horatio Alger–like stories of individuals such as Audie Murphy who pulled themselves "up by the bootstraps," success stories exemplified by the "rags to riches" theme, and patriotic stories of sacrifice, valor, service, and pride in a job well done.

In our Army they are the stories of today's heroes such as Medal of Honor recipient Sgt. 1st Class Paul Ray Smith and Sgt. Leigh Ann Hester, the first female soldier since World War II to receive the Silver Star medal for valor in combat. These are also the stories of wounded warriors who continue to persevere, of families who support their soldier's dedication and sacrifice, and ultimately of all soldiers who choose to selflessly serve their country.

Americans are renowned for their patriotism and their pioneering American spirit, which never gives in and never gives up. A deep respect for both rugged individualism and teamwork has been in existence since before the first cowboys, but that pioneering spirit has been responsible for the building of American folk heroes, and the spirit of the cowboy lives today in the spirit of the American soldier.

Americans hold soldiers in high regard. The Harris poll regarding "Confidence in Leaders of Major Institutions," released 28 February 2008, found that confidence in American institutions has declined since last year but that confidence in the American military has risen. In the survey for 2009, the military was again in first place, with 51 percent of Americans having a great deal of confidence, and small business was second, with 47 percent. Near the bottom of the poll was the press, with only about 10 percent of Americans expressing confidence in the media.[5]

Recently the mass media and the entertainment industry have skewed these iconic stories, often denigrating them with caustic wit and a debilitating skepticism, resulting in an inability to accept that truth, power, and goodness do still exist. "Reality television," in particular, has distorted—even, at times, perverted— the American success epic. From the business apprentice to the survivor on an island, the message is that to win you need to make someone else lose, turning life into a contest and conquest in which winners cheat and lie, criticize their opponents, and then revel in their failure.

Even the judges in these competitions can be demeaning and often vicious in their judgments—denigrating not only the contestants' efforts but often appearing to delight in crushing their egos and ultimately their sense of worth as human beings. There is no room for the cowboy or the soldier in these contests. Why do Americans watch these perversions of their values? An individual is raised up on a pedestal and then systematically knocked down. News reports covering the antics of out-of-control movie stars and those who qualify as celebrities only through their inherited wealth imply that their lifestyles are taken seriously. It is hard to imagine any of these contestants living each day with courage or taking pride in their work.

When a soldier falls from the pedestal, whether from a failure to uphold the values of the profession of arms or a more serious moral or ethical failure, the consequences appear even more stark. From a greater height of respect and expectation, the fall is that much greater, the judgment that much more harsh.

Even so the image of the soldier as hero remains strong and enduring. Author Ben Stein has commented on this matter of respect numerous times in the past several years. On his website he notes, "This country could last forever without the billionaire movie and TV stars in the magazines. We could not last a month without the men and women who fight for us. It's high time we got our priorities straight. Those guys and gals in Bagram and Ramadi and Fallujah and everywhere else, alive or dead or wounded, are the real stars, the ones who light up the night of tyranny with the light of freedom. We would not have a July 4th worth having without them. God bless them today and every day."[6]

S.Sgt. Jason Fetty is the first Army Reserve soldier to receive the Silver Star for heroism. A pharmacist from Parkersburg, West Virginia, Fetty volunteered for service with a Provincial Reconstruction Team (PRT) in Afghanistan. In February 2007 he found himself at the grand opening of a new medical facility in Khost, a significant and symbolic event denoting the tremendous progress made in the region. Medical professionals and government officials along with members of the PRT were there to celebrate the new facility when Fetty found himself faced with a man in a white lab coat who was behaving suspiciously. He quickly realized the man was a suicide bomber intent on disrupting the ceremony and destroying the facility itself. Fetty lured the bomber away from the crowd and stopped the attack.

There was a moment when Fetty realized that taking action was up to him and that there were really no other options. "You can feel it," he said. "You can see it. It's a general sinking feeling that things are not going to go right. You feel it in your gut."[7] In that moment he accepted the fact that he would do what had to be done.

Jason Fetty didn't try to become a hero. He didn't plan to. He did the right thing, making the only choice there was to be made, in an impossibly tough situation. I met Staff Sergeant Fetty at the Silver Star award ceremony in the Pentagon last November. He was characteristically modest and insisted that any other soldier in the same position would have done what he did. Fetty said to me that he was afraid that all the attention focused on him would detract from what really happened: "But I understand now that it doesn't take away from what happened, not really. That all of this doesn't change what really went on that day or take away from the bond between me and everyone in my unit." Fetty understood the need for the award and the subsequent media attention. "People need to know what happened," he said.

The discussion of basic soldier values needs to continue. Maj. Gen. Alan Bell, deputy commander of the U.S. Army Reserve Command, emphasizes knowledge of the Soldiers' Creed and asks soldiers about it whenever he conducts unit visits. "We need to keep a focus on the basic values," he said in a discussion about leadership. "Too many times people forget the basics and we see bad decisions and bad judgment as a result. I continue to remind commanders of the need for ethical decision making. It's all about modeling the behavior we expect of all who have the honor of wearing this uniform."[8]

The reminders to be effective, though, must be both constant and consistent from the chain of command. Externally the support must be likewise positive and unwavering. If Americans learned one thing from the Vietnam conflict it was the difference between disagreeing politically with the war and still maintaining support for the individuals who serve in uniform. American businessmen who give up their seats on flights so that soldiers returning from Iraq or Afghanistan can rest on their long trip home from the war zone are admirable. It is likewise heartwarming to see yellow ribbon magnets on SUVs, hear stories of community welcome-home ceremonies for returning National Guard and Reserve units, and learn of media coverage of soldier heroes or schoolchildren who write to soldiers or send them care packages.

Ben Stein, however, considers these expressions of support, while important, to be less than enough:

> Today and every day, men and women are fighting in Iraq in horrible conditions, with saboteurs and terrorists among them, to give that poor nation a chance to live in peace and democracy and to deny it as a haven for terrorism.
>
> How much do we owe them? Far, far more than we can ever pay them. How much do we owe them for spending Christmas so far from their families, so far from safety, so far from comfort? How much do we owe men and women who offer up their very lives for

*total strangers . . . ? . . . At this season of peace, all glory to the men
and women who go to war far from home so we can live and breathe
and get our feet massaged and pray in peace. We are nothing . . .
without them.*[9]

The soldier is America's cowboy in the twenty-first century, a role model of service and ethical behavior. He and she are volunteers, rugged individuals and members of a team, proudly affiliated with a values-based institution. They represent the American spirit.

9

The Cable, the Dish, and the Blog
Media Communication Ethics

*Congress shall make no law respecting an establishment of religion, or prohibiting
the free exercise thereof, or abridging the freedom of speech, or of the press;
or the right of the people to peaceably assemble, and to petition
the Government for a redress of grievances.*[1]

• • •

In the past ten years freedom of the press has been threatened not by external pressures but by internal abuses of the First Amendment, from news organizations with obvious political bias (an overly conservative or liberal bent) to the more than a dozen well-respected news organizations charged with ethical failures ranging from a failure to corroborate sources to blatant lies and outright plagiarism. While government organizations and the corporate world continually face questions of propaganda and attempts at press manipulation, the lapses in judgment by the mainstream media are just as egregious and more troubling for our culture.

The American education system plays a role in this failure as well, for its limited and traditionally indifferent role in requiring a strong ethical basis for journalism and mass communications students. The question is not how can this be fixed but, rather, is it too late for credibility to be restored to an institution repeatedly damaged by the excesses of its representatives? And, finally, does it matter? A new generation is turning away from the mainstream media and mass communications culture to the world of blogs and buzz, to smaller news sources fed by the cable and the satellite dish, to known trusted and reliable sources, their immediate circle of friends and family.

In her book *To Be One of Us: Cultural Conflict, Creative Democracy and Education*, Nancy Wareheim says that the ideological conflicts that characterize liberal education today are a microcosm of those in the greater American society.[2] This "crisis in humanities" education is perhaps no better illustrated than by the crisis in journalism and mass communications higher education. Journalism, with its inherent struggle for objectivity both within and outside of culture, its ability to politicize and influence issues—including those of public policy making, government decision making, and cultural agenda setting, renders the practices of mass communications and news reporting an ideal case in point.

The History

Philosophers throughout history have been concerned with the basic notion that communication shapes ideas. Seventeenth-century philosopher John Locke's "Essay Concerning Human Understanding" outlined the relationship between the development of the human mind and communication. He believed the mind is a blank slate at birth upon which all learning is inscribed by experience.[3] Communication, whether speech, art, or the written word, then serves not only as the delivery system for much of that developmental experience but also as the carrier of all learning, history, religion, and culture. We first learn from our parents, and then very quickly that experience expands to pictures, the printed word, and mass communications, especially television and, now, the Internet and its spawn.

At the beginning of the twentieth century, the concept of mass communications as a product of the industrial revolution began to assume an even greater role in American society. Walter Lippmann observed that people in the growing democracy naturally had a limited ability to observe important events firsthand and relied on the commercial news media to provide them with information. He viewed the press as a vehicle to provide us with views of the world from which we could "form pictures in our heads."[4]

Lippmann realized there often was a gap between reports of events and the reality of what had actually taken place. He also realized the importance of the process of selection in the coverage of certain events versus others. The significance of this act of selection was brought into focus in 1963, when Bernard Cohen made an observation now often repeated by media critics. He said, "The press may not be successful much of the time in telling people what to think but it is stunningly successful in telling people what to think about."[5]

The questions concerning the commercial media's agenda-setting function have been heard since its infancy—what is covered and how, along with questions concerning distortion or the accuracy of coverage distorted by a reporter's education, background, personal opinions, belief systems, and biases.

Since the first appearance of cheap, mass-circulation newspapers in America in the mid-1800s, improvements in technology, such as the invention of the web press (which printed simultaneously on both sides of large rolls of paper), population growth and increasing urbanization, as well as general literacy, served to increase the daily demand for information. Even as early as the 1890s, American newspapers faced a growing social issue, one brought about as a byproduct of their commercial success: "Reaching the masses meant lowering intellectual and cultural standards, appealing to the emotions, and adopting popular and sometimes radical, causes."[6]

However, there is a genuine sense of community and shared history in mass media coverage of events. Nearly every American who was alive at the time remembers where they were on November 22, 1963, when President John F. Kennedy was shot in Dallas, and where they were during the subsequent live television coverage of his state funeral. Others point to that July day in 1969 when they watched astronaut Neil Armstrong first set foot on the moon and heard those static-laden words "That's one small step for man, one giant leap for mankind." The "now" generation of digital natives had their own victory celebration, sharing in the broadcast moment in January 2009, when Barack Obama raised his right hand to take the oath of office as president.

Such events can be viewed as common cultural references, shared civic or social events, or even national "town hall meetings." Naturally there are few live news broadcasts of such truly national significance in our increasingly fragmented culture, but those we do watch as a nation contain incredibly powerful and even culturally defining images.

Images continue to demonstrate the power of the civilian news media to bring us stories of events both unifying and pluralistic, destructive or invigorating—to explain them, to analyze them, to define them, to place them in social and historical context, and even, if only briefly, to use them to center and integrate our culture. They also are highly susceptible to influence, agenda-setting priorities, and politicization.

A Social Responsibility

Criticism of the commercial news media's objectivity is likewise not a new phenomenon. Checkbook journalism (paying for stories) dates back at least to 1904, when the *New York Times* paid Admiral Peary four thousand dollars for the exclusive rights to his story of the trip to the North Pole. Today a whole genre of celebrity and personality magazines and their television clones, from *People* to *Entertainment News*, thrives on the American fascination with celebrities and even ordinary people who sell stories about their personal and not-so-private lives. This is the American version of having a royal family about whom to gossip.

The terms "muckraking" and "yellow journalism" may conjure up images ranging from Watergate to former President Clinton's liaison with an intern, but those terms originated with newspaper political coverage in the early twentieth century. Charges of political and corporate advertising influence on the objectivity of the news are likewise not new.[7]

The role of the press as cultural critic blossomed fully in the 1960s, moving rapidly across the civil and social strife of the divided American landscape, developing from the relatively benign agenda-setting function to full-blown mainstream status as a political and moral influencer. This occurred at the height of the Vietnam War. With its often startling and up-close images of human destruction and suffering brought to the living room in real time or close to it, television news came into its prime as well, competing directly with the newspaper for the attention of the American people.

The media has long been accused of political bias in news coverage, from the personal to political. A recent UCLA study succeeded in proving such bias is real, revealing that of the twenty media outlets studied, eighteeen scored left of center, with the *Wall Street Journal* in first place followed by the *CBS Evening News*, *New York Times*, and *Los Angeles Times*. At the other end of the spectrum, Fox News' *Special Report with Brit Hume* and the *Washington Times* scored right of the average U.S. voter. The most centrist program proved to be the *NewsHour with Jim Lehrer*. CNN's *News Night with Aaron Brown* and ABC's *Good Morning America* followed.[8]

The Power and the Glory

In their 1991 book *Unreliable Sources: A Guide to Detecting Bias in News Media*, authors Martin A. Lee and Norman Solomon wryly observed, "Popular legends assure us the American press is committed above all to seeking and speaking the truth . . . no matter who might be offended."[9] While sociologists have expressed doubts the news media have the power to directly change people's attitudes, the media do provide immense quantities of information upon which people base attitudes and develop them.[10] Agenda setting is a legitimate function of the press, not only in a societal or cultural sense but also in a very real political sense.

Given the logical role of the press as the "gatekeepers" of information in our society, it seems the education and training of journalists and editors would receive great attention and scrutiny, that there naturally would be particular attention paid to the development and enforcement of ethical standards and the necessity for fullness and objectivity in reporting. Interestingly, even given the current number of books on the market describing the pervasive lack of media objectivity, the presumed tendency to rush toward judgment, the overbearing influence of advertising and a number of other ills, there appears to be little mainstream discussion of the increasing necessity for a focus on ethical standards in journalism education.

Even recent scandals such as the *New York Times'* tale of the lies told by junior reporter Jayson Blair; "Rathergate," in which CBS News fired four staffers for their role in preparing a false story about President George W. Bush's National Guard service; and *USA Today's* 2005 firing of Pentagon news correspondent Tom

Squitieri for plagiarism have done little to hold the public's attention for even a full weekly news cycle. This is somewhat surprising given that these and other national-level reporters who have fallen from grace represent some of the most respected and prestigious news institutions in America. Less surprising is the fact *USA Today* offered no formal apology to those offended or disadvantaged by their stories.

Conversely the November 2005 discovery that U.S. military contractors in Iraq paid Iraqi newspapers to print news stories resulted in a full-scale media attack on what was termed "government propaganda." Reporters, editors, and pundits alike scrambled to be the latest and the loudest to criticize the military's attempt to influence audiences, the implication being that American audiences were likewise being targeted by those involved. It appears there is little media appetite for intro-spection or self-criticism but an almost unspoken bond to respond with one voice to any attempt to "control the message" from or by government organizations.

Larry DiRita, at the time a spokesman for the Pentagon, ordered a full-scale investigation into the affair. The inquiry discovered no wrongdoing on the part of the Lincoln group but did bring to the forefront the burgeoning issue of mili-tary communications and information operations. As Thom Shanker reported in the *Washington Post*, "The question for the Pentagon is its proper role in shaping perceptions abroad. Particularly in a modern world connected by satellite televi-sion and the Internet, misleading information and lies could easily migrate into American news outlets, as could the perception that false information is being spread by the Pentagon."[11]

Political and government organizations are not without blame. Many have contributed their own efforts to attempt to spin, obscure, and deflect reports of their own mismanagement or overt partisan political efforts, fueling suspicion that attempts to "control the message" are confirmation that the government has plenty it is trying to hide from public examination and discussion. Yet it is that innate suspicion, that immediate presumption of guilt, that often sours public opinion when mass media seek to out corporate or government wrongdoing. The approach is distasteful, and the media often suffer from a negative response just as they seek to provoke righteous outrage or expose wrongdoing in others.

Several years ago the Army averted yet another major media "sting" when CNN researchers in New York were caught in the act of conducting an under-cover operation. The staffers were seen in several New York City recruiting stations, wearing strange-looking yellow glasses (meant to conceal recording and videotap-ing devices) and asking about opportunities to enlist in the U.S. Army as well as commenting on how those opportunities would be affected by the course of the war. Yet when one staff member's odd behavior resulted in a recruiter pressing for more details about his interest, the staffer ran out of the station and toward the street. Recruiters watched in open-mouthed amazement as he opened the door to a nearby minivan, exposing large amounts of broadcasting and recording gear inside. Word spread quickly across the Army's Recruiting Command about the undercover sting conducted by a traditionally well-respected network.

Army Public Affairs staffers contacted CNN that afternoon. While producers and the CNN ethics counselor were chagrined at having been caught, they likewise were defensive about the validity about their chosen means of gathering information, the Army's open door policy and reputation for transparency notwithstanding.

These cautionary tales can and should apply equally to the ethics of both the subjects of news coverage as well as those who do the reporting. Less-than-forthright means of gathering information are as questionable as the ethics of those who seek to influence the news gatherers. Even so it appears there is little media appetite for introspection or self-criticism, but an almost unspoken bond does exist within the media to respond with one voice to any and all perceived First Amendment threats or attempts to "control the message" by government organizations.

Tools of the Trade

Perhaps one longstanding reason for the lack of public discussion regarding media ethics is that journalism is not a true profession. Unlike law, medicine, and engineering, for example, there are no set standards for admission to the field. Demonstrated mastery of a specialized field of knowledge is not required. Likewise there is no governing body to oversee the ethical application of the tools of the trade, to accredit or license practitioners. There is no oath or sworn promise to uphold ethical principles. Indeed, practitioners make every effort to close their fraternity to a select few with "approved" educations and standard ethics, freezing out those who have different backgrounds and outlooks and different perceptions of what news is and how it should be conveyed.

Throughout the historical development and evolution of the profession, at least in anglophone nations, journalists have largely displaced scholars as the "public explainers," those who put into context society's events, politics, culture, and public policies. Yet they are not accountable in a political sense, and as many journalists decry the move to "civic journalism," they often claim no responsibility to society either.[12] What should the entrance requirements be for those who would assume such a role? If we regard journalism more as a trade than a profession, there still should be benchmarks for qualifications to serve at the journeyman level and standards for practice of the craft. How can the layman judge if tradecraft is well executed? Obviously affiliation with a "big name" such as the *New York Times* or CBS is no guarantee of quality.

In 1996 the Society of Professional Journalists (SPJ) voluntarily enacted a new code of ethics "after months of study and debate among the society's members."[13] Yet even the existence of such a code appears hypocritical. How can it be enforced if not through the willing acquiescence of its volunteer membership in an occupation where the greatest majority are not joiners, willing to submit themselves to standards and positions set by others in the group? Without a common educational core, how can there be consensus as to the meaning and importance of these often-unenforceable guidelines? They cannot be called rules.

As for the significance of ethical instruction, the SPJ website provides a number of ethics case studies for journalism instructors and others interested in the issue. As the site mildly understates, "There seems to be no shortage of ethical issues in journalism these days."[14] As they enter the field, college journalism graduates are not required to have any systematic education or training in history, philosophy, political systems, liberal arts, natural sciences, or sociological and economic analysis.[15] There are no core standards. In addition it is possible and indeed likely, given the increasing fragmentation of our society and audiences, for an individual to enter the field of journalism without a journalism degree, or in fact without any degree.

Freelance writers and bloggers need no credentials, and the explosion of individual sites in the blogosphere gives testimony to the expansion of unique voices, each telling a unique story. Each person's site demonstrates that the lowest common denominator in communications—that is, the ability to read and write and a willingness to serve as witness and recorder to history—can attract an audience.

According to E. D. Hirsch Jr. in *Cultural Literacy: What Every American Needs to Know,* literate culture is the "common currency" for social and economic exchange in our democracy and the sole available ticket to full citizenship.[16] Yet what we are seeing is a trend moving away from any recognizable need for a common currency and toward an increasingly fragmented, specialized, and isolated number of spheres. We no longer trust mainstream journalism to interpret the national experience, and statistics tell us that the world of blogs is becoming the source of news for a whole new generation.

In fact a recent study by Edelman/Technorati shows that 34 percent of all bloggers gain recognition as authorities in their fields. Thirty-two percent see themselves as diarists, the witnesses to history who then record it. There are still others who serve as media watchdogs, daily correcting and discussing the errors they see in the mainstream media. Their popularity is obviously increasing since, according to the 2005 Edelman Public Relations trust barometer, "'regular people' are three times more credible than established figures of authority."[17]

Richard Edelman, CEO of Edelman Public Relations, said, "Even as technologies are enabling the democratization of communications, public trust in major institutions is eroding. This steady wearing down, a result of the deluge of scandals in the traditional power centers, touches every facet of society, involving business, government and the media."[18]

The Role of Education

Perhaps not even experienced practitioners recognize the significance of the education issue. Charles Peters of the *Washington Monthly* spoke several years ago to a group of young people considering careers in journalism. He was asked what preparation he would recommend for students who hoped to be reporters: "Peters launched into a tirade about the importance of studying history, literature, government, science, or *anything* other than journalism. It was very easy to learn journalism on the job, but it was very hard to make up for the lost general education that the college years should represent."[19]

What should such a general education program include? E. D. Hirsch argues in *Cultural Literacy* that a humanistic education based on Plato's principles is not necessarily conservative. Whether we agree or disagree, as do Nancy Wareheim and many of her fellow critics, certainly there should at least be a basic body of knowledge on which to agree or disagree.

It cannot be as disingenuously simple as authors Lee and Solomon claim in *Unreliable Sources*: "Aspiring journalists get started by imitating established journalists and most of the careers that follow consist largely of echoing others in the profession."[20] It defies logic that such a pattern could produce objective, insightful journalists—ones unaffected by the marketplace or pull of advertising, personal prejudice, politics, or liberal or conservative bias—or even reporters unable to resist the ease and convenience of copying or imitating others rather than making the effort to learn and think independently. If the goal is to support a fourth estate that informs a broad public about relevant issues, events, and facts so that citizens can make informed decisions, then education must play a decisive role in preparing members of that fourth estate to assume this responsibility. It is the place where we must begin.

Education Basics

A recent nationwide study of high school journalism sponsored by the Freedom Forum Foundation Media Center revealed a number of interesting facts about journalism curricula, teachers, students, and their products. In New York State, for example, where journalism teachers are not required to be certified, if is fairly common to see teachers without a remote connection to journalism advising the student newspaper. Many high school journalism teachers have had no journalism education themselves, and most are responsible for developing their own curricula.

Some experts attribute the shortage of college-educated high school journalism teachers to the small number of colleges offering majors in journalism education. A 1991 survey of nearly eight hundred journalism educators showed the majority first got involved with journalism when assigned to teach it.[21] Tom Rolnicki, executive director of the National Scholastic Press Association, believes most high school journalism teachers typically use "old texts and things they can photocopy." Surveys show only 18 percent use textbooks, and those who do are not closely tied to them.[22]

The most popular high school texts of the last twenty years usually have chapters on news writing, editing, interviewing, reporting, layout and design, advertising, business management, and other tasks. More recent texts tend to stress student press rights and responsibilities and focus more on technological advancements such as desktop publishing, broadcasting, and other elements of mass communications and broadcast media. Typically there is little room for the history and theory of mass communications and culture. The 1997 edition of *Rolnicki's Scholastic Journalism* was the first to include a chapter on journalistic ethics.[23]

One chapter is clearly insufficient for the topic, given the sophistication and complexity of ethical issues journalists face today. In contrast there is a growing education focus on legal matters in many curricula. This issue has assumed even greater importance as professional journalists must be concerned with complex issues such as libel, the Freedom of Information Act, protection of sources, privacy issues, and others, which, apart from ethical concerns, are extremely important to the well-being of the profession itself. While they may not appear in some textbooks, short essays on these topics are included in the journalist's "bible," *The Associated Press Stylebook*.[24] Related topics include those concerns that could form the basis of lawsuits or result in legal rulings restraining the efforts of the press.

Educational Implications

Should there be one "best" system for educating young journalists at the high school level as well as college? Perhaps this question should be better phrased: Should there even be a system? As a starting point, if you accept the idea of a system, there should be a "common core" of studies. Such a core curriculum should seek to do more than train students in the tasks and tools of the tradecraft of journalism, whether print or broadcast. It should seek to educate young reporters in a sound foundation on the purpose and responsibilities of journalism as well as its status as a carrier of culture. Students also should learn the power of the agenda-setting function, issues associated with the selection of news topics and the ability to weigh and represent all sides of a question. Ethics must be emphasized and their importance repeated.

The Accrediting Council on Education in Journalism and Mass Communications (ACEJMC) accredits 106 American college and university programs in journalism and mass communications. Ethical principles, the ability to "demonstrate an understanding of professional ethical principles and work ethically in pursuit of truth, accuracy, fairness and diversity," is one element evaluated by the council as part of its review of an institution's curriculum.[25]

It is patently obvious that one single element is insufficient in aiding budding professionals as they construct a strong moral and ethical compass; the ethical standard for instruction and the core ethics curriculum must be expanded. Journalism and mass communications students must be able to distinguish bias in existing news agencies and organizations, reporters, and editors, and must recognize the creep of editorial viewpoints into straight news and critically examine and edit their own work for evidence of bias and personal prejudice. These capabilities are essential to maintaining a strong ethic in a society ever more pressured by the marketplace and measured by the polls.

It is likewise important to distinguish between legal rights and responsibilities under the First Amendment and the larger moral sense of responsibility to society. Development of this more general understanding is critical to the education of those who would seek to become "explainers."

Whether we call them "explainers," "watchdogs," or even "guardians," it is an incontrovertible fact that this is the role of journalism in our society. Plato's *Republic* discussed the theme of how a society could best be reshaped so as to provide optimum opportunities for its citizens to discover the best in themselves from various angles. In the chapters discussing the Guardians (rulers), he says for them to have the best chance of success, they "must have the right education, whatever that may be."[26]

Today's self-styled guardians of democracy are not rulers, but as journalists, they often interpret both the acts and words of leaders and rulers, as well as develop and promote trends. Their task is to explain change to an increasingly fast-paced, complex, and technologically advanced world. That alone makes their education a matter of tremendous social importance.

We cannot ignore the need for ethical education in a profession clearly in the throes of a long, spiraling decline. For nearly twenty years national opinion polls have repeatedly shown the civilian news media is widely distrusted as an American institution.[27] While all institutions have slipped in the public eye, it is perhaps the news media that has fallen the furthest from grace. In 2005 Americans again placed the U.S. military in first place as America's most trusted institution with a 74 percent approval rating, while the news media, both print and broadcast, rated near the bottom tier of the poll, with a meager 28 percent score.

The cultural side effects of this crisis may be even more far-reaching than we suspect: "The coverage we see . . . aggravates today's prevailing despair and cynicism about public life."[28] As a result mainstream media continues to lose market share. Not surprisingly newspaper readership continues to drop, and the most popular national television news program in the 18-to-34-year-old group is one that makes fun of the news itself, Jon Stewart's *Daily Show* on the Comedy Channel.

There is little doubt Americans think journalists are sloppier, less professional, less caring, less honest about their mistakes, more biased, and generally more harmful to democracy than they were in the 1980s. James Fallows, in examining the connection between news and democratic government, said, "If they held themselves as responsible for the rise of public cynicism as they hold 'venal' politicians and the 'selfish' public; if they considered that the license they have to criticize and defame comes with an implied responsibility to serve the public—if they did all or any of these things, they would make journalism more useful, public life stronger, and themselves far more worthy of esteem."[29] This is a strong indictment of the American news media, and there are choices to be made as a result. We must continue to try to reform the pattern of inadequate and unethical journalism and mass communications and develop a professional core and a new paradigm for ethical standards that can withstand the most intense scrutiny.

If we are unable to break this cycle, the trend toward a slow and painful death for the mainstream media will continue. The falling circulation that has plagued U.S. newspapers over the past ten years is likely to continue (see fig. 9-1). The few papers that saw circulation increases had specific outreach programs and strategies for gaining new readers. Those that work to promote circulation are struggling to maintain rather than increase their numbers. News magazines are in an even greater state of crisis. By the end of February 2005, *Time* magazine was pursuing another round of buyouts to encourage early staff retirements while analysts continued to predict the imminent death *of U.S. News and World Report* and *Newsweek*.

A factor that should not be discounted is the unwillingness of too large a part of the population to not concern itself with important public issues, using the cable, the dish, and the blog to turn away from the central core of community news and democratic debate of the day, preferring to be selectively entertained rather than broadly informed. All the training and all the ethics that can be inculcated will mean nothing if too few are interested in buying the product.

For newspapers to succeed on the Web they must compete with online outlets and cable and either fend off or merge successfully with broadcast incursions into the Web as well. This year newspapers are expected to expand their Internet presence via more interactive features and online video and audio options.[30] But is it enough or a case of too little too late? The decline of newspapers may well be

Print and Online Newspaper Readership

Read yesterday ...	2006	2008	2010	06-10 change
Any newspaper*	43	39	37	-6
In print	38	30	26	-12
Online	9	13	17	+8
Print only	34	25	21	-13
Online only	5	9	11	+6
Both print & online	4	5	5	+1

PEW RESEARCH CENTER June 8–28, 2010. Q9,11, 20.

Figures may not add exactly to subtotals because of rounding.

*Includes respondents who reported reading a newspaper yesterday as well as those who said they got news online yesterday and, when prompted, said they visited the websites of one or more newspapers when online (Q20).

FIGURE 9-1. Results of the Pew Research Center's poll "Print and Online Newspaper Readership," 8–28 June 2010

inevitable, leaving the venerable evening newscast to face its own fall upon the departure of its dinosaurs—Jennings, Rather, and their likes.

This competition will continue to intensify online, the beginnings of it spurred by eyewitness news direct from the front lines in near real time, delivered via the growing number of soldier bloggers in the Iraq and Afghanistan war zones: "It is sometimes said that journalism is the first draft of history. Military bloggers are now writing the first draft of war. Soldier blogs share real stories. . . . They are the most embedded reporters of all."[31]

At the Crossroads

When he stepped down as president of the Accrediting Council in 2004, Jerry Crepos said in a departing address, "No, I don't have a solution and I'm delighted that it is no longer my job to grapple with one. But I do believe that we could develop a set of outcomes that we hope for from ethics and fairness instruction, a procedure that wouldn't be prescriptive. Perhaps we even should publish a guide to meeting new standards on ethics education, as we did on diversity last year; it was full of ideas, not requirements."[32]

Following his departure the council's new committee sought in the summer of 2005 to gather information on best practices in teaching journalism ethics. Those best practices must be shared widely, implemented fully, and expanded. And, yes, they should be prescriptive. It is time. Otherwise journalism is in very real danger of becoming the road less taken, for the prospects for a successful career as a journalist are wavering now.

Finally, this change must occur quickly. While education is not the complete answer to the problem of ethics, it does represent a necessary beginning. The rapid pace of technological change in information delivery systems is about to accelerate with an explosive move. Just as the printing press ignited the renaissance, computers and the Internet have increased the number and speed of information delivery systems, providing us with access to a breadth and depth of information unimaginable in an earlier age.

Technology is exploding so quickly that author Ian Jukes says we have its development in "dog years," that one year of Internet development is the same as seven in any other medium.[33] This has profound implications for the future of education and the direction of learning. The question is not how to get ahead of the coming wave but how, if it isn't already too late, to keep up with it. A Nielsen Company review of media trends in December 2010 revealed social networks/blogs now account for one in every four and a half minutes online (almost six hours), with the average visitor spending 66 percent more time on the Web than he or she did last year (three hours and thirty-one minutes).

A system of solid ethical grounding should be able to withstand whatever new developments offer in the way of technology, speed, or spread of information. There just needs to be a system.

10

Challenge to Change: Developing Leaders for the Twenty-First Century

Strategic communication is vital to U.S. national security. It is an increasingly powerful, multi-dimensional instrument that is critical to America's interests and to achieving the nation's strategic goals.[1]

• • •

A Week at Fort Knox[2]

The general officers and senior colonels juggled their coffee cups and blackberries, mingling in the hallway just inside the deputy assistant secretariat's office suite. Several stood at the window, watching a light snow slowly drift down on the frozen Kentucky landscape. The electronic sign was already flashing at the door: "Board in Session! No Admittance!" Even though it was a few years old, on 22 January 2015, the Human Resource Command's Headquarters building at Fort Knox, Kentucky, still looked brand new.

There were fifteen board members in total, all carefully selected based on their background, experience, and knowledge of the varied applications of information in Military Support to Public Diplomacy (MPD), Civil Affairs Information, and Military Information Support Operations (MISO; formerly Psychological Operations), and public and community affairs. The board members represented the related career fields included in the young cohort, commonly referred to as the Combined Information Team (CIT). Still other board members represented the combat arms branches and were selected to represent the Combined Arms Team.

All of the CIT board members also had extensive command experience, and many were experienced in leading complex, multinational, strategic communications planning efforts, operations, and integrated joint and combined campaigns. All were there to serve on the 2015 Non-Kinetic Effects Command and Key Billet Selection Board. They had traveled to Fort Knox over the weekend, most from within CONUS, but also from as far away as U.S. Army Africa Headquarters in Vincenza, Italy, and theaters of operations in the Middle East and South America. One colonel had arrived late on Sunday due to flight delays in Argentina.

Just after 8:00 a.m. on Monday, the two-week board convened and members took the oath required of all Army Selection Board members. The Human Resource Command (HRC) staff began their program of opening briefings on the board processes. The board president, Major General Goodfriend, opened with the following comments:

> Ladies and gentlemen, I know you are excited to have the privilege of sitting this board. The transformation of our Army's information enterprise has enabled us to break down the stovepipes and silos that previously existed and move forward rapidly with the integration of broad and effective information programs. For many years, we focused too narrowly on terminologies and the capabilities of our equipment and not enough on the innovation and creativity of our people. Career information officers were focused on hardware and information delivery systems rather than on the act of crafting information products with intent and design. Frankly, I never cared whether communication had an s on it or not. It never was about the machine that delivers information.

The members laughed as they recalled the artificiality of that old construct. "It has been a hard road to get to this point," Goodfriend reminded them:

> Internally, there was terrific resistance to developing the capabilities of these related career fields and giving them the resources necessary for success. We had to force change within the Army.[3] Externally we've had to deal with significant opposition in the mainstream media, though not so much from the general public. It took a massive public education effort and, as you may recall, early exposure of those unsuccessful terror attacks in Chicago several years ago to open the door to a greater understanding of the power of strategic communication. Only in the past three years have the White House, State Department, and Department of Defense been able to move beyond old urban myths and fears of propaganda and news manipulation to the new and level information playing field where we stand today.
>
> What did Americans abhor? What did we fear? Being manipulated . . . or being influenced? After all, we are influenced every day, and the complexity of dealing with all of the challenges we face in a loud, crowded, and dangerous information battle space, whether at home or abroad, made coordination difficult and cooperation among the varied elements of our joint force nearly impossible.

The board recorders exchanged telling glances. Several of the majors were well aware that Goodfriend was renowned for his gift of rhetoric. He liked to talk, and the story of America's effort to manage its information capabilities was a favorite subject of his. Once on a roll, he would recite the entire history of the fight to win hearts and minds, at home and abroad, while along the way citing every government advance and setback of the past ten years to manage Information Operations, (IO), Public Affairs (PA), and MISO in a generally recognized and accepted manner. Once he began to discuss the Office of Strategic Influence under Secretary of Defense Rumsfeld, the recorders settled in, knowing he was good for at least twenty minutes more of discourse on that particular chapter in communications history.

Later several recognized the end of the presentation was approaching as Goodfriend's tone shifted and his pace changed from that of the classroom lecture to a discussion of definitions of the career fields:

> *The definitions of each are on your board reference cards. They are clear and unambiguous. Truth is the foundation of all of our communication efforts, whether in a press conference, brochure, blog entry, tweet, or community meeting. All elements of our combined information team work together to prevent, to deter, and to fight to achieve victory. Let me remind you that no one career field or specialty is the "right" road to success. In some cases an IO officer may not have had sufficient broadening experiences to compete successfully for some of these IO or SC positions. In other cases you may think a PAO is not ready for command. Pay careful attention to the senior rater comments; these and your own best judgment should be your guide.*

One of the junior board members, a brigadier general representing Field Artillery, interjected: "Sir, there are still comments about how IO leeches into public affairs and undermines our credibility in dealing with foreign audiences. How can Strategic Communications as a capstone concept ever separate—"

Goodfriend held up a hand and interrupted the general:

> *Let me give you an example you can understand. Let's say a Maneuver Enhancement Brigade commander is an Aviation Branch officer. He has at least three subordinate battalions. One is Engineer, one is Military Police (MP), and one is Chemical. Do you think the Engineer unit tries to do the mission of the MP unit? Does the Chemical unit attempt to do the Explosive Ordnance Detachment (EOD) mission? The same is true of Information Operations and Public Affairs. Their capabilities are different. Complimentary, yes. But different. This is exactly why we now teach much more about the nonkinetic fight in our professional schools, from the Basic Officer's Leadership Course (BOLC) forward and in every part of the NCO education system.*

Goodfriend paused, looked around the table, then continued: "You all have the latest version of DA Pam 600-3 on your desktop and version four of FM 3-0 for reference. The CSA's memorandum of guidance is likewise your guide as we select these senior officers and civilians for command and key billets in the Departments of Army, DoD, and State. I expect that you will see varied backgrounds and experiences with no clear-cut career pattern that can be identified as a 'standard.' That is a good thing. These senior leaders are different from those like us, who still recall being told we needed to stay on only the operational path, to be an S-3 or an XO, to succeed. With that, ladies and gentlemen, let's get to work." (See fig. 10-1.)

Board members nodded their agreement. They were there to determine the fully qualified and best-qualified Army colonels (year groups 1990–92) to be named to the more than thirty different commands and key billet positions available in fiscal year 2016. There were commands available in Civil Affairs, MISO, and Information Operations. For the first time this board would likewise consider Public Affairs colonels for the Public Affairs functional commands in all Army components. All functional areas also had selected key billets they had carefully designated over the past several years for priority fill.

Finally, the board also would select colonels for the available key Strategic Communications positions in the unified commands and joint task force orga-

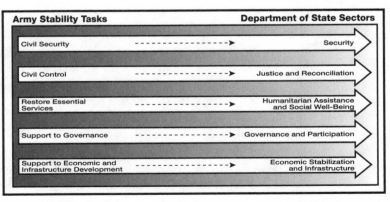

FIGURE 10-1. A "Whole of Government" Approach
—*Army Field Manual 3-0, Information Operations*

nizations worldwide. Qualified officers from any of the above fields were eligible to compete. They were aware of the careful timing associated with each position turnover and the Senior Leader Development Office (SLDO) requirement for every colonel selected for command or a key billet at this level to attend the Non-Kinetic Commander's Pre-Command Course (NKCPCC) at the DoD Center for Communications Excellence at Fort Meade, Maryland.

The center will be established at Fort Meade in 2013 to take advantage of the aggregate experience and capabilities of the various commands at Fort Meade. These include the Asymmetric Warfare Center, which served as an integrator for the Psychological Operations and Information Operations portion of the course, along with the First IO Command; DoD's Defense Information School (DINFOS), which taught the Strategic Communications, messaging and message integration, and professional ethics portions of the courses; and the Army Reserve's First Public Affairs Support Element (PASE) Command, which taught how to apply and integrate PA aspects of information to the overall information campaign.

As often happens during board meetings, the staff had briefed board members on Monday that there would be some "special boards" for smaller groups of members to vote on later in the week. On Thursday, after voting on the main body of candidates had ended and the staff was reviewing the tally for any ties that had to be broken and for system accuracy, the fifteen board members found that they had been broken into several smaller groups based on their basic branches or specialized experience. The recorders instructed each group separately on their additional responsibilities.

Special Boards

One of those special boards, composed solely of general officer members, was charged with reviewing the files of those public affairs officers who would be departing command or key billets in the summer of 2015 to determine the best officers to nominate for the key joint PA billets in OSD (PA) and in the unified commands and with the State Department. The unified command PA director positions were all flag officers (07), and the Navy had occupied the majority of these jobs since their inception in 2013. While the jobs were flag billets, they were actually brevet promotions. An officer going to one of these jobs for a three-year tenure would have the option of retiring at flag rank at the end of the tour or reverting to the grade of 06 in order to continue to serve. Many thought that Army officers had a very good chance of capturing at least four of the seven positions coming open in 2016.

Certainly those who were competitive for the positions believed that holding this joint job would place them in good stead for consideration for the position at the top of the Army's PA field, the chief of public affairs. Beyond that there were two new two-star PA positions in DoD pending approval for fiscal year 2016 and a two-star position as commander of the Defense Media Activity Command (DMAC) and PDASD (PA). The DoD positions were open to all services to nominate their flag officers for consideration. Other specialty boards were likewise considering Information Operations colonels for select positions in the J-39 staffs of the unified commands and those with embassy or state department experience for key billets on J-5 staffs and within the offices of the political advisors of the unified commands.

Finally there were also several special boards to select high-performing Department of the Army civilians for certain designated positions[4] determined key to developing strong Senior Executive Service (SES) candidates. Serving Army SESs also participated in these boards as directed by the Civilian Senior Leader Management Office (CSLMO). The civilians competing on these boards were all volunteers for the fast track; nearly all had deployment experience and were resident senior service college graduates. They had all signed mobility agreements at the beginning of their careers, and the Army's Civilian Talent Management Office (CTMO) managed their careers and schooling. There was particularly tough competition for the senior-level Public Affairs positions, and all of the civilians competing had signed mobility statements and agreements for short-term developmental assignments as available. They represented a young and aggressive career civilian population. Since the mass departure of career Army civilians in 2011 (popularly referred to as the BRAC Massacre),[5] the civilian component of the Army was still rebuilding. The civilian force of 2015, however, was young, aggressive, innovative, and terrifically committed to the ideal of government service.

A Joint Staff initiative was likewise beginning to mature. A number of general officer billets were interchangeable with SES members, permitting the best-qualified individual to be selected. This effort, initiated with a pilot study in 2013, was a terrific morale booster for the senior civilian leadership throughout the department.

Three of the most visible positions in DoD's Strategic Communications field included the position of chief of staff to the DASD (PD) in USD (P), director of Counter Terror Information Planning (CTIP) within the office of the ASD for Homeland Defense (HD), director of strategic outreach for the Secretary of the Army, and director of strategic communications in OSD (PA). Previously one had been filled by career diplomats from the State Department. Other coveted positions included deputy director of strategic communications at the National Security Council (NSC), communications director at the Counter Terror Center (CTC), and the new communications director positions added to every ambassador's country team in over twenty-five countries whose development had been designated as vital to America's interests.

Information Tasks

"In modern conflict, information has become as important as lethal action in determining the outcome of operations. Every engagement, battle, and major operation requires complementary information operations to both inform a global audience and to influence audiences within the operational area . . ."

FM 3-0, Chapter 4

Task	Information Engagement	Command and Control Warfare	Information Protection	Operations Security	Military Deception
Intended Effects	• Inform and educate internal and external publics • Influence the behavior of target audiences	• Degrade, disrupt, destroy, and exploit enemy command and control	• Protect friendly computer networks and communication means	• Deny vital intelligence on friendly forces to hostile collection	• Confuse enemy decision-makers

FIGURE 10-2. The Importance of Information

Public Affairs Career Field Redesign, 2013–14

There was a major emphasis on giving serious consideration to developing the non-kinetic fight, beginning with the current administration's focus on fiscal restraint for military spending in 2009. In some areas this translated into a major redesign of several career fields. While information operations had developed into a robust Army capability during the years 2004–7, public affairs appeared to lag behind. The specialty itself had become even narrower during the preceding decade and produced officers who often appeared more rigidly identified with the function of conducting media relations and writing sluggish press releases than with overall coordination of internal and public information activities.

The redesign:

- PA functional commands were established in each component to improve training readiness and mission alignment with the DoD and Army. Because the size of the force had long meant PA units were "under the radar," the change also facilitated command and control, equipping, and manning.

- The number and capabilities of Army PA NCOs were increased to enable them to have leadership and career broadening opportunities. The NCO corps had been decimated in the 1980s and had never recovered from the grade cuts.

- Ten additional CSM positions were established in the PA field to support the new commands and provide senior NCO leadership to the field. The force structure bill payer for this and the NCO corps growth was the U.S. Army Field Band. The band was quietly disestablished in 2010, and its absence went largely unremarked across the Army.

- Career opportunities and talent development and management for Army civilian public affairs professionals were enhanced.

- The Public Affairs Branch was enhanced to increase its capabilities and to standardize the training and education opportunities for Army soldiers, NCOs, and officers to make them commensurate with those of the other services.

- Career paths were modified to develop more broadly capable public affairs officers, competitive with those who served in the information operations field.

- The Army's communications graduate degree program was standardized to bring it into line with the other services. This was done to enable Army PAOs and civilians to obtain a one-year graduate degree in either mass communications management or strategic communication. The Navy was designated to serve as executive agent for officer education, based on its success at San Diego State University. NCOs could likewise pursue a one-year graduate degree in convergence communications or new media. The Air Force was selected to manage enlisted education for the services and based its programs on the success at Syracuse University.

- Emphasis on professional training with industry programs was restored, as a follow-on to deployment experiences, at the grades of 04/SFC/YA-3 in all components.

- The proponent for Combat Camera was moved from the Army Signal Corps to the Office of the Chief of Public Affairs, bringing it structurally in line with how the other services operate. Command and operational oversight for Combat Camera operations in all services transferred to the Defense Media Activity Command in 2011.

- The two public affairs enlisted specialties were merged into one, a general communications specialist who could write, speak, photograph, record, and transmit information using varied multimedia. This merger simplified training in convergence communications and strengthened skills in new media.[6]

- Course curriculum in the PA, MISO, IO, CA, and FAO fields were changed to include a basic awareness and understanding of the capabilities of each of the other areas as complimentary and supporting elements of the non-kinetic fight.

The Way Ahead

Friday morning at 10:00 a.m., HRC commander Major General Goodfriend entered the boardroom to conduct the summary briefing and adjourn the board. All of the board members expressed their confidence in the board process and in the fairness of the selections and results. In open discussion they commented on several other points they wanted to emphasize to the Chief of Staff and Secretary of the Army. First, they could see that establishment of the Public Affairs Branch and the career field redesign was bearing fruit. Officers were broadly skilled, deeply experienced, and extremely competitive for the available communications enterprise command and key billet positions. Several remarked on the integration of civilian talent into the pool of senior leaders as a major factor in increasing overall career field professionalism and as a testament to the diverse capabilities of the civilian work force. A final point was delivered by the board president:

> *I was most reassured to see how the increase in professional education on the nonkinetic fight continues to pay benefits for the entire communications enterprise, and really for our Army as a whole. The internal competition is virtually nil, and the cross-fertilization between those with attaché experience to those who have experience in State or tactical IO operations, or with the Armed Forces Network news, . . . well, it is coming together. I am most encouraged.*
>
> *The strong network of our communicators and their abilities as a team to shape a public discourse about our Army is a testament to their creativity, ability to operate comfortably with a variety of social media, and willingness to work together as a combined information team. I read about some truly remarkable achievements in the evaluation reports of these officers and civilians. I am confident we are moving forward.*

Public Affairs Colonel-Level Commands and Key Billets[7]

- First Public Affairs Command (USAR)
- Second Public Affairs Command (AC)
- Third Public Affairs Command (ARNG)
- Commander, Joint Public Affairs Support Element (Joint, nominative)
- Commander, Defense Information School (Joint)
- Director, Army Public Affairs Center
- Commander, Defense Media Activity Command (Army Element)
- Director of Media Relations, OSD (PA) (Joint, nominative)

Strategic Communication Colonel-Level Commands and Key Billets[8]

- Director of Strategic Outreach, Office of the Chief of Staff, Army
- Chief of Staff to the DASD (PD)
- NSC Deputy Director of Communications
- CTC Director of Communication
- Unified Command Deputy Director of SC

2016 Acronyms

ACCP	Army Communications Enterprise Positions
AFRICOM	U.S. Africa Command
ASD (HD)	Assistant Secretary of Defense for Homeland Defense
BRAC	Base Realignment and Closure
CIT	Combined Information Team
CMF	Career Management Field
CONUS	Continental United States
CSLMO	Civilian Senior Leader Management Office
CTIP	Counter Terror Information Planning
CTMO	Civilian Talent Management Office
DASD (PA)	Deputy Assistant Secretary of Defense for Public Affairs
DASD (PD)	Deputy Assistant Secretary of Defense for Public Diplomacy
DINFOS	Defense Information School
DMAC	Defense Media Activity Command
FAO	Foreign Area Officer
HRC	Human Resources Command
IO	Information Operations
JECC	Joint Enabling Capabilities Command
MISO	Military Information Support Operations
MPD	Military Support to Public Diplomacy
NKC-PCC	Non-Kinetic Commander's Pre-Command Course
NKE	Non-Kinetic Effects
OSD	Office of the Secretary of Defense
PDASD (PA)	Principal Deputy Assistant Secretary of Defense for Public Affairs
PSYOPS	Psychological Operations
SES	Senior Executive Service
SLDO	Senior Leader Development Office
USD (P)	Under Secretary of Defense for Policy

2015 Combined Information Team Career Definitions

Strategic Communication (DoD). Focused U.S. government processes and efforts to understand and engage key audiences to create, strengthen, or preserve conditions favorable to advance national interests and objectives through use of coordinated information, themes, plans, programs, and actions synchronized with other elements.

Public Affairs (Joint). Those public information, command information, and community relations activities directed toward both the external and internal publics with interest in the Department of Defense.

Information Operations (DoD). The integrated employment of the core capabilities of Electronic Warfare, Computer Network Operations, Psychological Operations, and Military Deception and Operations Security in concert with specified supporting and related capabilities to influence, disrupt, corrupt, or usurp adversarial human and automated decision making while protecting our own.

Military Information Support Operations (formerly known as Psychological Operations) (DoD). Planned operations to convey selected information and indicators to influence the emotions, motives, objective reasoning, and, ultimately, the behavior of foreign governments, organizations, groups, and individuals. The purpose of psychological operations is to induce or reinforce foreign attitudes and behavior favorable to the originator's objectives.

Appendix I

Fitzwater's Ten Commandments

The following is from the January 1989 issue of *U.S. Army Public Affairs Monthly Update*, a professional development bulletin published by the Army's Office of the Chief of Public Affairs and distributed to Public Affairs offices throughout the army. The advice it contains, from Marlin Fitzwater, assistant to the president and press secretary, hung on my office wall for many years.

A Government Spokesman Shall Not

Lie. *"A good press secretary doesn't have to. There are always times when you have to say, 'I can say nothing for national security reasons,' or 'I can't tell you anything for diplomatic reasons,' or 'I can't tell you anything because we're still working on a decision that hasn't come out yet.' There are ticklish times when you want to run out the back door, but I don't think there are any reasons to lie."*

Restrict the Public's Right to Know. *"I instinctively know that this is the people's right. Government is set up to do for people what they can't do for themselves. I don't know of any limitations; there are always limitations in terms of the ability of the government to provide information, but I wouldn't limit the other side of the equation."*

Lose Sight of Who's Master. *"The first master has to be the person you work for, in my case the President. You cannot lose sight of that, it's like a rudder. Reporters are not monolithic in their needs for information or in the way they view information."*

Be Intimidated by the Bureaucracy. *"You're always pushing the bureaucracy a little bit, because the press is pushing you. . . . A spokesman is called upon to give more information than the people on the inside think you have to give. So you have to get it yourself and trust your judgment."*

Forget to Ask Questions. *"The old joke about feeling an elephant in the dark and deciding it's a camel is true. . . . You're always having to hustle, having to go to the meetings, having to call people, having to weigh all the information. Then a spokesman has to make judgments about the information he receives."*

Take Anything for Granted. *"You get the right to participate by virtue of having sense and having someone higher up give [that right] to you, but it doesn't go with the job."*

Relinquish Access. *"There are so many people who come to a problem from different angles. The diplomat sees an issue one way, the military genius sees it from a different perspective, the politician sees it from a third. That's why in this job access to the President is so important. The President's feeling is the only right one."*

Lose a Sense of Humor. *"It's incredibly important in the sense of being able to see the irony of things, to laugh at oneself and deal with problems. . . . On the other hand, nothing gets you in trouble faster than a good quip. As just about every press secretary in history will attest, it's usually their jokes they would like to take back most."*

Hate the Press. *"I like reporters, I like these people. That's the basic element of any relationship. You get mad at individuals, just like anybody else."*

Hold out on the Press. *"I always found that if you answer the phone and try to answer the questions then you're 90 percent of the way there. . . . I have a lot of sympathy for reporters sitting outside the government and trying to find out what in the hell is going on."*

Appendix II

Department of Defense
Principles of Information

The Principles of Information constitute the underlying public affairs philosophy for Defense.gov and the Department of Defense. The principles are codified as enclosure (2) to Department of Defense Directive 5122.5 of 27 September 2000.

> It is Department of Defense policy to make available timely and accurate information so that the public, the Congress, and the news media may assess and understand the facts about national security and defense strategy. Requests for information from organizations and private citizens shall be answered quickly. In carrying out that DoD policy, the following principles of information shall apply:
>
> Information shall be made fully and readily available, consistent with statutory requirements, unless its release is precluded by national security constraints or valid statutory mandates or exceptions. The Freedom of Information Act will be supported in both letter and spirit.
>
> A free flow of general and military information shall be made available, without censorship or propaganda, to the men and women of the Armed Forces and their dependents.
>
> Information will not be classified or otherwise withheld to protect the Government from criticism or embarrassment.
>
> Information shall be withheld when disclosure would adversely affect national security, threaten the safety or privacy of U.S. Government personnel or their families, violate the privacy of the citizens of the United States, or be contrary to law.
>
> The Department of Defense's obligation to provide the public with information on DoD major programs may require detailed Public Affairs (PA) planning and coordination in the Department of Defense and with the other Government Agencies. Such activity is to expedite the flow of information to the public; propaganda has no place in DoD public affairs programs.
>
> The Assistant Secretary of Defense for Public Affairs has the primary responsibility for carrying out the commitment represented by these Principles.

Notes

Chapter 1. Military Media Relations

This article first appeared in the *Wright Stuff* 5, no. 18 (26 August 2010), published online by the U.S. Air Force Air University, http://www.au.af.mil/au/aunews/.

1. From informal discussions and interviews with nearly five thousand officers during media relations training sessions conducted from 1992 to 2009. Sessions were held at the U.S. Army Logistics Management College, George C. Marshall European Center for Security Studies, NATO School, Sweden's International Training Command, and Norway's Command and Staff Training Center.

2. Department of Defense, "Secretary of Defense Memorandum: Interaction with the Media," 2 July 2010, http://www.fas.org/sgp/othergov/dod/media.pdf/.

3. Office of the Secretary of Defense (Public Affairs), *Data Depot*, 22 July 2010.

4. Gallup poll results, 22 July 2010.

5. Joint Chiefs of Staff, *Doctrine for Public Affairs in Joint Operations*, Joint Publication 3-61, 14 May 1997.

6. Warren P. Strobel, *Late-Breaking Foreign Policy: The News Media's Influence on Peace Operations* (Washington, D.C.: U.S. Institute of Peace Press, 1997), 219.

Chapter 2. Crisis Communication and Conflict Storytelling

1. Sajjan M. Gohel, director for international security, Asia Pacific Foundation, "New Media and Ideology: The Battle for the 'Message' in the Information Age," presentation given at the George C. Marshall European Center for Security Studies conference "Exploring Dimensions in Countering Ideological Support for Terrorism," Amman, Jordan, 29 September 2009.

2. Pew Research Center, "The State of the News Media, an Annual Report on American Journalism," Pew Research Center's Project for Excellence in Journalism 2010, http://stateofthemedia.org/2011/.

3. Ibid.

4. Eric Dezenhall, *Nail 'Em! Confronting High-Profile Attacks on Celebrities and Businesses* (Amherst, N.Y.: Prometheus Books, 2003).

Chapter 3. Strategic Communication and the Battle of Ideas

This article is reprinted with the permission of the U.S. Naval War College, International Law Studies Department. It first appeared in *International Law Studies*, vol. 83, *Global Legal Challenges: Command of the Commons, Strategic Communications and Natural Disasters*.

1. Secretary of Defense Donald Rumsfeld in remarks at the Council on Foreign Relations, 17 February 2006, New York, N.Y.

2. Karen Hughes, remarks given at the Baker Institute for Public Policy, Houston, Tex., 29 March 2006.

3. Ibid.

4. Rumsfeld at the Council on Foreign Relations.

5. James G. Stavridis, *Partnership for the Americas: Western Hemisphere Strategy and U.S. Southern Command* (Washington, D.C.: National Defense University Press, 2010), 250.

6. Department of Defense, Terms of Reference, Deputy Assistant Secretary of Defense (Joint Communication) in the Office of the Assistant Secretary of Defense (Public Affairs), Approved by Mr. Lawrence DiRita, Assistant Secretary of Defense (Public Affairs), 6 January 2005.

7. Quadrennial Defense Review Execution Roadmap, draft, February 2006.

8. Frank Aukofer and William P. Lawrence, eds., *America's Team: The Odd Couple—A Report on the Relationship Between the Media and the Military* (Nashville: Freedom Forum First Amendment Center at Vanderbilt University, 1995).

9. Richard Halloran, "Soldiers and Scribblers Revisited: Working with the Media," *Parameters* 21, no. 1 (Spring 1991): 10–20.

10. Linda Robinson, "The Propaganda War," *U.S. News and World Report*, 29 May 2006, 29–31.

11. Aukofer and Lawrence, *America's Team*.

12. Sydney Freedberg Jr., "Daily Briefing," *National Journal*, 17 February 2006.

13. Ibid.

14. Statement of Frank Thorp, deputy assistant secretary of defense (Joint Communication) at a closed door hearing of the House Armed Services Subcommittee on Terrorism, Unconventional Threats and Capabilities, 109th Cong., 19 July 2006.

15. Freedberg, "Daily Briefing."

Chapter 4. Toward Strategic Communication

This article is reprinted with the permission of *Military Review: The Professional Journal of the U.S. Army*, Combined Arms Center, Fort Leavenworth, Kansas. It was originally published in the July–August 2007 issue of *Military Review*.

1. Department of State, U.S. National Strategy for Public Diplomacy and Strategic Communication, summary of conclusions, memo.

2. Department of the Army, *Counterinsurgency*, FM 3-24, December 2006, 1–3.

3. Robert D. Heinl, *Dictionary of Military and Naval Quotations* (Annapolis: Naval Institute Press, 1966), 175.

4. Bredberg in *Sydsvenskan*.

5. Al-Khazin, "Commander in Chief," 20.

6. Quadrennial Defense Review Working Group.

7. Hughes, remarks given at the Baker Institute for Public Policy.

8. Ibid.

9. Rumsfeld at the Council on Foreign Relations.

10. Department of Defense, Terms of Reference.

11. Quadrennial Defense Review Execution Roadmap.

12. Department of the Army, objective template for "Enhanced Strategic Communication," internal working document.

13. Maj. Gen. Patrick H. Brady, "Telling the Army Story: 'As It Is, Not As It Should Be,'" *Army* magazine, September 1990.

14. U.S. Army Infantry School, BCT lesson 071-D-2390.

15. Brian Bender, "Pentagon Studying Its War Errors," *Boston Globe*, 16 August 2006, 1.

16. Aukofer and Lawrence, *America's Team*.

17. Robinson, "Propaganda War," 29–31.

Chapter 5. The Missing Element: Strategic Communication

This article first appeared in the February 2009 issue of *Proceedings* magazine. It is reprinted from *Proceedings* with permission; copyright © 2009 U.S. Naval Institute.

1. F. Scott Fitzgerald, *The Great Gatsby* (1925; reprint, New York: Charles Scribner's Sons, 1953), 182.

2. George C. Marshall European Center for Security Studies, *Advancing International Cooperation in Countering Ideological Support for Terrorism*

(CIST): *Toward Building a Comprehensive Strategy*, report, 2007, 17, http://www.marshallcenter.org/.

3. Lawrence Pintak, Jeremy Ginges, and Nicholas Felton, "Views of Arab Journalists," *New York Times*, 25 May 2008.

4. Tony Blankley and Oliver Horn, "Strategizing Strategic Communication," WebMemo #1939, 29 May 2008, http://www.heritage.org/Research/NationalSecurity/wm1939.cfm/.

5. Philip Seib, J.D., "The Al-Qaeda Media Machine," *Military Review*, May–June 2008.

6. I was an Army advisor to the DSB Strategic Communications Study and participated in many of the sessions.

7. Office of the Secretary of Defense, *Report of the Defense Science Board Task Force on Strategic Communication*, January 2008.

8. Ibid.

9. Ibid., 115.

10. Nick Gillespie and Matt Welch, "How Dallas Won the Cold War," *Washington Post*, 27 April 2008, B02.

11. Jeffrey Jones, "Strategic Communications: A Mandate for the United States," *Joint Forces Quarterly*, No. 39, http://www.dtic.mil/doctrine/jel/jfq_pubs/1839.ct/.

12. Blankley and Horn, "Strategizing Strategic Communication."

13. George C. Marshall, "Marshall Plan speech," Harvard University, 5 June 1947.

Chapter 6. The Golden Thread: Building Champions and Connections

1. Committee on Trauma, *Advanced Trauma Life Support Program for Doctors* (Chicago: American College of Surgeons, 2008).

2. Center for Strategic and International Studies, *CSIS Commission on Smart Power: A Smarter More Secure America*, co-chairs Richard L. Armitage and Joseph Nye Jr. (Washington, D.C.: Center for Strategic and International Studies, 2007), 49–50.

3. Chip Heath and Dan Heath, *Made to Stick: Why Some Ideas Survive and Others Die* (New York: Random House, 2007), 13.

4. Center for Strategic and International Studies, *CSIS Commission*, 33.

5. Justin Peters, "Trust Falls," *Columbia Journalism Review*, 4 March 2010.

6. There are approximately one hundred of these volunteer ambassadors who serve as the representatives of the chief, Army Reserve and as advocates for soldiers and families in communities across the country.

7. Rebecca Murga, "Full Circle," *Warrior Citizen*, Fall 2010, 27.

8. eMarketer, "The Continued Rise of Blogging," released 23 September 2010.

9. Pew Research Center, "Americans Spending More Time Following the News," released 12 September 2010.

10. *Travel and Leisure*, October 2010, 70–71.

11. John F. Kennedy, quoted in the *Boston Globe*'s magazine, the *Week*, 19 November 2010.

Chapter 7. The Toxic Information Environment

This article first appeared in the *Wright Stuff* 5, no. 13 (24 June 2010), published online by the U.S. Air Force Air University, http://www.au.af.mil/au/aunews/.

1. John D. Sutter, "Ashton Kutcher Challenges CNN to Twitter Popularity Contest," CNN, 15 April 2009, http://www.cnn.com/2009/TECH/04/15/ashton.cnn.twitter.battle/.

2. Sam Tanenhaus, "North Star: Populism, Politics and the Power of Sarah Palin," *New Yorker*, 7 December 2009, 84–89.

3. Daniel Dayan and Elihu Katz, *Media Events* (Cambridge: Harvard University Press, 1992).

4. Tanenhaus, "North Star," 86.

5. Pew Research Center, "State of the News Media."

6. Brian Williams, quoted in *Newsweek*, 7 December 2007, 20.

7. Jonah Goldberg in the *Los Angeles Times* as quoted in the *Week*, 25 December 2009–8 January 2010, 14.

8. Harris poll, "Confidence in Leaders of Institutions," 10 March 2010.

9. Despair, Inc., http://www.despair.com/.

10. Nicholas A. Christakis and James H. Fowler, *Connected: The Surprising Power of Our Social Networks and How They Shape Our Lives* (New York: Little, Brown, 2009), 275.

11. President Barack Obama, speech, United Nations General Assembly, 23 September 2009.

Chapter 8. The Cowboy and the Soldier: Military Communication Ethics

Reprinted from *Army* magazine, copyright © 2008 and with permission of the Association of the U.S. Army.

1. Remarks by Secretary of the Army Pete Geren, 10 April 2007, Washington, D.C.

2. "The Code of the West," in *Cowboy Ethics: What Wall Street Can Learn from the Code of the West*, by James P. Owen (Ketchum, Idaho: Stoecklein Publishing and Photography, 2004), 24.

3. U.S. Army poster.

4. Ben Stein, *How Successful People Win: Using "Bunkhouse Logic" to Get What You Want in Life* (Carlsbad, Calif.: Hay House, 1981), 3.

5. Harris poll, "Confidence in Leaders of Major Institutions."

6. Ben Stein, "Stuff Ben Wrote," 4 August 2001, BenStein.com, http://www.benstein.com/070404/.

7. Donna Miles, "First Reserve Soldier Receives Silver Star for Valor in Afghanistan," 15 October 2007, American Forces Information Service.

8. Maj. Gen. Alan Bell, deputy commander of the U.S. Army Reserve Command, conversation with the author, 6 March 2008.

9. Ben Stein, *The Real Stars: In Today's America, Who Are the Real Heroes?* (Carlsbad, Calif.: New Beginnings Press, 2007), 144.

Chapter 9. The Cable, the Dish, and the Blog: Media Communication Ethics

1. The Constitution of the United States.

2. Nancy Wareheim, *To Be One of Us: Cultural Conflict, Creative Democracy, and Education* (Albany: State University of New York Press, 1993), xxiv.

3. Shearon A. Lowery and Melvin L. DeFleur, *Milestones in Mass Communication Research: Media Effects* (New York: Longman, 1983), 328.

4. Ibid.

5. Ibid., 329.

6. John A. Garraty, *The American Nation: A History of the United States since 1865* (New York: Times Books, 1993), 165.

7. Howard Kurtz, *Media Circus: The Trouble with America's Newspapers* (New York: Times Books, 1993), 358.

8. "Media Bias Is Real, Finds UCLA Political Scientist," UCLA press release, 14 December 2005.

9. Martin A. Lee and Norman Solomon, *Unreliable Sources: A Guide to Detecting Bias in News Media* (New York: Carol, 1990), xiii.

10. Lowery and DeFleur, *Milestones in Mass Communication Research*, 329.

11. Thom Shanker, "Propaganda," *Washington Post*, 21 March 2006.

12. James Fallows, *Breaking the News: How the Media Undermine American Democracy* (New York: Pantheon Books, 1996), 150.

13. Society of Professional Journalists, http://www.spj.org/ethics.asp/.

14. Ibid.

15. Fallows, *Breaking the News*, 150.

16. E. D. Hirsch Jr., *Cultural Literacy: What Every American Needs to Know* (Boston: Houghton Mifflin, 1987), 21–22.

17. Edelman Public Relations, https://extranet.edelman.com/bloggerstudy/.

18. Richard Edelman, CEO Edelman Public Relations, "Public RelationSHIPS and Communications in the Age of Personal Media," http://www.edelman.com/.

19. Freedom Forum, *Death by Cheeseburger: High School Journalism in the 1990s and Beyond* (Nashville: Freedom Forum Foundation, 1994), 12–15.

20. Lee and Solomon, *Unreliable Sources*, 15–16.

21. Freedom Forum, *Death by Cheeseburger*, 17.

22. Ibid.

23. Ibid.

24. *The Associated Press Stylebook*, ed. N. Goldstein (New York: Associated Press, 1993).

25. ACEJMC Accrediting Standards, Standard 2, "Curriculum and Instruction."

26. *The Republic of Plato*, trans. with notes by Francis MacDonald Cornford (London: Oxford University Press, 1945), 108.

27. Lydia Saad, "Military Again Tops 'Confidence in Institutions' List," Gallup Organization, 1 June 2005.

28. Mike Thorp, "The Media's New Fix," *U.S. News and World Report*, 18 March 1996, 32.

29. Fallows, *Breaking the News*, 65.

30. MediaWeek.com, http://www.mediaweek.com/mediaweek/images/pdf/forecast/.

31. Joseph Mallia, "Tangled in the Web," *Long Island Newsday*, 3 January 2006, 4.

32. Jerry Crepos, "Remarks on Stepping Down as President of the Accrediting Council on Education in Journalism and Mass Communications," address to members of the council, 30 April 2004, Cambridge, Mass.

33. Ian Jukes, "From Gutenberg to Gates to Google and Beyond: Education for the Online World," presentation, 14 June 2005, http://www.committedsardine.com/.

Chapter 10. Challenge to Change: Developing Leaders for the Twenty-First Century

This article first appeared on the Army War College Strategic Studies website in the section called "Of Interest" on 15 June 2009. It has since been updated.

1. Office of the Secretary of Defense, *Report of the Defense Science Board Task Force.*

2. The scenario that follows, and the Human Resource Command Board, is a fictional construct, developed to illustrate how an integrated-systems approach to building a communications enterprise could fully leverage and connect the Army's information capabilities. This is a vision for one potential approach.

3. A summary of the changes made within the public affairs career field follows.

4. Twenty positions were designated as key Army Communications Enterprise Positions (ACEP) billets in 2012.

5. The "BRAC Massacre," also known as the "BRAC Attack," refers to the exodus of thousands of Army civilians from the rolls as the physical moves required by the Base Realignment and Closure Act of 2005 occurred in 2010–11. On 30 June 2010, termed "Black Friday," more than 25,000 Army civilians left government service, swelling the retirement rolls.

6. The changes in combat camera and the merger of enlisted specialties reduced the training burden on the Defense Information School and permitted the curricula to support all services equally. This freed up intellectual capital, and following accreditation in 2010, the school's Interactive New Media Department became a state-of-the-art reference in university mass communications programs nationwide.

7. Not all inclusive.

8. Not all inclusive. More continue to be developed as the information career field develops.

Bibliography

What follows is a list of some of the best references in my personal communications library. I have read and reread many of these books and articles, referred to them repeatedly over the years, and used them in my own writing and teaching efforts. Others have been significant for the ideas and the varied points of view they present. Some of these sources may no longer be in print, such as several books written by former White House spokespersons. However, the gems of advice they include are priceless, and searching these volumes out is worthwhile.

Crisis Communications

Dezenhall, Eric. *Nail 'Em! Confronting High-Profile Attacks on Celebrities and Businesses*. Amherst, N.Y.: Prometheus Books, 2003.

Gowing, Nik. *"Skyful of Lies" and Black Swans: The New Tyranny of Shifting Information Power in Crises*. Oxford: Oxford University Press, 2009.

Pinsdorf, Marion K. *Communicating When Your Company Is Under Siege: Surviving Public Crisis*. Lexington, Mass: Lexington Press, 1987.

Journalism and Broadcasting

Bates, Stephen. *If No News Send Rumors: Anecdotes of American Journalism*. New York: Henry Holt, 1989.

Cohen, Elliot D. *Philosophical Issues in Journalism*. New York: Oxford University Press, 1992.

Donovan, Robert J., and Ray Scherer. *Unsilent Revolution: Television News and American Public Life*. New York: Cambridge University Press, 1992.

Gillmor, Dan. *We the Media: Grassroots Journalism by the People, for the People*. Sebastopol, Calif.: O'Reilly Media, 2006.

Hamill, Pete. *News Is a Verb: Journalism at the End of the Twentieth Century*. New York: Ballantine, 1998.

Starr, Paul. *The Creation of the Media: Political Origins of Modern Communications*. New York: Basic Books, 2004.

Marketing

Brown, Paul B. *Marketing Masters: Lessons in the Art of Marketing*. New York: Harper & Row, 1988.

Gladwell, Malcolm. *Outliers: The Story of Success*. New York: Little, Brown, 2008.

Godin, Seth. *Permission Marketing*. New York: Simon & Schuster, 1999.

Salzman, Marian, Ira Matathia, and Ann O'Reilly. *Buzz: Harness the Power of Influence and Create Demand*. Hoboken, N.J.: John Wiley & Sons, 2003.

Scott, David Meerman. *The New Rules of Marketing and Public Relations: How to Use News Releases, Blogs, Podcasting, Viral Marketing and Online Media to Reach Buyers Directly*. Hoboken, N.J.: John Wiley & Sons, 2007.

Media Criticism

Downie, Leonard, Jr., and Robert G. Kaiser. *The News about the News: American Journalism in Peril*. New York: Vintage Books, 2003.

Fallows, James. *Breaking the News: How the Media Undermine American Democracy*. New York: Pantheon Books, 1996.

Glasser, Theodore L., and Howard E. Sypher. *The Journalism of Outrage: Investigative Reporting and Agenda Building in America*. New York: Guilford Press, 1991.

Goldberg, Bernard. *Arrogance: Rescuing America from the Media Elite*. New York: Warner Books, 2003.

———. *Bias: A CBS Insider Exposes How the Media Distort the News*. Washington, D.C.: Regnery Press, 2002.

Kovach, Bill, and Tom Rosenstiel. *Warp Speed: America in the Age of Mixed Media*. New York: Century Foundation Press, 1999.

Kurtz, Howard. *Media Circus: The Trouble with American Newspapers*. New York: Random House, 1993.

———. *Spin Cycle: Inside the Clinton Propaganda Machine*. New York: Free Press, 1998.

Lee, Martin A., and Norman Solomon. *Unreliable Sources: A Guide to Detecting Bias in News Media*. New York: Carol, 1991.

Sanford, Bruce W. *Don't Shoot the Messenger: How Our Growing Hatred of the Media Threatens Free Speech for All of Us*. New York: Free Press, 1999.

Solomon, Norman. *The Habits of Highly Deceptive Media: Decoding Spin and Lies in Mainstream News*. Monroe, Maine: Common Courage Press, 1999.

Media Law

Burrell, Hedley, ed. *An Unfettered Press.* Reston, Va.: Center for Foreign
    Journalists, 1992.
Holsinger, Ralph. *Media Law.* New York: Random House, 1987.
Schoenfeld, Gabriel. *Necessary Secrets: National Security, The Media, and the Rule
    of Law.* New York: W. W. Norton, 2010.

Military References

Caton, Jeffrey L., Blane R. Clark, Jeffrey L. Groh, and Dennis Murphy, eds.
    *Information as Power: An Anthology of Selected U.S. Army War College
    Student Papers.* Vol. 3. Carlisle, Pa.: U.S. Army War College, 2009.
NATO Military Policy on Public Information. MC 457. September 2001.
U.S. Department of the Army. *Operations.* FM 3-0. Chap. 7, "Information
    Superiority."
U.S. Department of Defense. *Army Public Affairs Handbook.* ST 45-07-01. 5
    April 2007.
U.S. Departments of the Army and the Air Force, National Guard Bureau. *Public
    Affairs Guidance on National Guard Bureau Environmental Programs.*
    Washington, D.C.: National Guard Bureau, n.d.
U.S. Joint Chiefs of Staff. *Doctrine for Public Affairs in Joint Operations.* Joint
    Publication 3-61. 14 May 1997.
U.S. Joint Forces Command, Joint Warfighting Center. *Commander's Handbook
    for Strategic Communication and Communication Strategy.* Version 3.0.
    24 June 2010.

New Media

Drudge, Matt, with Julia Phillips. *The Drudge Manifesto.* New York: New
    American Library, 2000.
Godin, Seth. *Linchpin.* New York, Simon & Schuster, 2010.
Jarvis, Jeff. *What Would Google Do.* New York: HarperCollins, 2009.
Jenkins, Henry. *Convergence Culture: Where Old and New Media Collide.* New
    York: New York University Press, 2006.
Levine, Rick, Christopher Locke, Doc Searls, David Weinberger. *The Cluetrain
    Manifesto.* New York: Basic Books, 2001.
Livingston, Geoff, with Brian Solis. *Now Is Gone: A Primer on New Media for
    Executives and Entrepreneurs.* Baltimore: Bartleby Press, 2007.
Shapiro, Andrew L. *The Control Revolution: How the Internet Is Putting Individuals
    in Charge and Changing the World We Know.* New York: Perseus Books,
    1999.

Propaganda

Bettinghaus, Erwin P., and Michael J. Cody. *Persuasive Communication.* New York: Holt, Rinehart and Winston, 1987.

Goldstein, Col. Frank L., USAF, and Col. Benjamin F. Findley Jr., USAFR. *Psychological Operations Principles and Case Studies.* Maxwell Air Force Base, Ala.: Air University Press, 1996.

Pincus, Walter. "Pentagon May Have Mixed Propaganda with PR." *Washington Post,* 12 December 2008, 2.

Thomson, Oliver. *Easily Led: A History of Propaganda.* Phoenix Mill, UK: Sutton Publishing, 1999.

Public Diplomacy

Pincus, Walter. "GAO Calls for a New Priority on Public Diplomacy." *Washington Post,* 12 January 2009, 11.

Tuch, Hans N. *Communicating with the World: U.S. Public Diplomacy Overseas.* New York: St. Martin's Press, 1990.

Public Relations

Culligan, Matthew J., and Dolph Greene. *Getting Back to the Basics of Public Relations and Publicity.* New York: Crown, 1982.

Cutlip, Scott M., Allen H. Center, and Glen M. Broome. *Effective Public Relations.* 9th ed. New York: Prentice Hall, 2005.

Doorley, John, and Helio Fred Garcia. *Reputation Management: The Key to Successful Public Relations and Corporate Communication.* New York: Taylor & Francis, 2007.

Norris, James S. *Public Relations.* Englewood Cliffs, N.J.: Prentice Hall, 1984.

Oliver, Col. Keith, USMC (Ret.). *Command Attention: Promoting Your Organization the Marine Corps Way.* Annapolis: Naval Institute Press, 2009.

Pritchard, Robert S. "Effective Public Relations." In *Convergent Journalism: An Introduction,* by Robert S. Pritchard. New York: Focal Press, 2005.

Young, Davis. *Building Your Company's Good Name: How to Create and Protect the Reputation Your Organization Wants and Deserves.* New York: American Management Association, 1996.

Reports

Aukofer, Frank, and William P. Lawrence, eds. *America's Team: The Odd Couple—A Report on the Relationship Between the Media and the Military.* Nashville: Freedom Forum First Amendment Center at Vanderbilt University, 1995.

George C. Marshall European Center for Security Studies. *Advancing International Cooperation in Countering Ideological Support for Terrorism (CIST): Toward Building a Comprehensive Strategy.* Report. 2007. http://www.marshallcenter.org/.

Gowing, Nik. *Media Coverage: Help or Hindrance in Conflict Prevention.* New York: Carnegie Commission on Preventing Deadly Conflict, 1997.

Helmus, Todd C., Christopher Paul, and Russell W. Glenn, eds. *Enlisting Madison Avenue: The Marketing Approach to Earning Popular Support in Theaters of Operation.* Santa Monica, Calif.: Rand Corporation, 2007.

LaMay, Craig, Martha Fitzsimon, and Jeanne Sahadi, eds. *The Media at War: The Press and the Persian Gulf Conflict.* New York: Gannett Foundation Media Center, 1991.

McCormick Tribune Foundation. *The Military-Media Relationship.* 2006 and 2007. McCormick Tribune Foundation conference series. http://www.mccormickfoundation.org/.

Office of the Secretary of Defense. *Report of the Defense Science Board Task Force on Strategic Communication.* January 2008. Washington, D.C., January 2008.

Paul, Christopher, and James J. Kim, eds. *Reporters on the Battlefield: The Embedded Press System in Historical Context.* Santa Monica, Calif.: Rand Corporation, 2004.

Pew Research Center for the People and the Press. http://www.people-press.org/.

Reagan, Michael, ed. *CNN Reports: KATRINA—State of Emergency.* Atlanta: Lionheart Books, 2005.

Reuters Institute for the Study of Journalism. http://www.reutersinstitute.politics.ox.ac.uk/.

Spokespeople

Clarke, Torie. *Lipstick on a Pig: Winning in the No-Spin Era by Someone Who Knows the Game.* New York: Free Press, 2006.

Fitzwater, Marlin. *Call the Briefing: Reagan and Bush and Sam and Helen: A Decade with Presidents and the Press.* New York: Times Books, 1995.

Fleischer, Ari. *Taking Heat: The President, the Press, and My Years in the White House.* New York: HarperCollins, 2005.

Strategic Communication, Information Operations, and Public Diplomacy

Carsten, Michael D., ed. *International Law Studies.* Vol. 83, *Global Legal Challenges: Command of the Commons, Strategic Communications and Natural Disasters.* Newport, R.I.: Naval War College, 2007.

Corman, Steven R., Angela Trethewery, and H. L. Goodall Jr., eds. *Weapons of Mass Persuasion: Strategic Communication to Combat Violent Extremism.* New York: Peter Lang, 2008.

Davidson, Joel R. *Armchair Warriors: Private Citizens, Popular Press, and the Rise of American Power.* Annapolis: Naval Institute Press, 2008.

Forno, Richard, and Ronald Baklarz. *The Art of Information Warfare: Insight into the Knowledge Warrior Philosophy.* New York: Universal Publishers, 2007.

Stavridis, James G. *Partnership for the Americas: Western Hemisphere Strategy and U.S. Southern Command.* Washington, D.C.: National Defense University Press, 2010.

Tuch, Hans N. *Communicating with the World: U.S. Public Diplomacy Overseas.* New York: St. Martin's Press, 1990.

War Coverage and Peacekeeping

Fialka, John J. *Hotel Warriors: Covering the Gulf War.* Washington, D.C.: Woodrow Wilson Center Press, 1991.

Hammond, William M. *Public Affairs: The Military and the Media, 1962–1968.* Washington, D.C.: Center for Military History, 1988.

———. *Public Affairs: The Military and the Media, 1968–1973.* Washington, D.C.: Center for Military History, 1995.

———. *Reporting Vietnam: Media and Military at War.* Lawrence: University Press of Kansas, 1998.

Katovsky, Bill, and Timothy Carlsen. *The Media at War in Iraq: An Oral History.* Guilford, Conn.: Lyons Press, 2003.

Moeller, Susan D. *Compassion Fatigue: How the Media Sell Disease, Famine, War, and Death.* New York: Routledge, 1999.

Newman, Johanna. *Lights, Camera, War: Is Media Technology Driving International Politics?* New York: St. Martin's Press, 1996.

Pulwers, Jack E. *The Press of Battle: The G.I. Reporter and the American People.* Raleigh, N.C.: Ivy House, 2003.

Strobel, Warren P. *Late-Breaking Foreign Policy: The News Media Influence on Peace Operations.* Washington, D.C.: U.S. Institute of Peace Press, 1997.

Index

Page numbers followed by an *f* indicate figures.

About the Author

MARI K. EDER is a career public relations and communications professional with over 30 years of diverse communications experience with the Department of Defense. She is recognized internationally as an expert in strategic communication and has lectured and written extensively on the subject, both in the United States and Europe.